Computer Sound Synthesis
for the Electronic Musician

TECHNOLOGY
S e r i e s

Series introduction

The Focal Press Music Technology Series is intended to fill a
growing need for authoritative books to support college and
university courses in music technology, sound recording, multimedia and their related fields. The books will also be of value
to professionals already working in these areas and who want
either to update their knowledge or to familiarise themselves
with topics that have not been part of their mainstream occupa-
tions.

Information technology and digital systems are now widely
used in the production of sound and in the composition of music
for a wide range of end uses. Those working in these fields need
to understand the principles of sound, musical acoustics, sound
synthesis, digital audio, video and computer systems. This is a
tall order, but people with this breadth of knowledge are increasingly sought after by employers. The series will explain the
technology and techniques in a manner which is both readable
and factually concise, avoiding the chattiness, informality and
technical woolliness of many books on music technology. The
authors are all experts in their fields and many come from teach-
ing and research backgrounds.

Dr Francis Rumsey
Series Editor

Titles in the Series

The Audio Workstation Handbook   *Francis Rumsey*

Acoustics and Psychoacoustics   *David Howard and James Angus*

Sound and Recording: an introduction (3rd edition)
Francis Rumsey

Sound Synthesis and Sampling (book and CDROM)
Martin Russ

MIDI Systems and Control (2nd edition)   *Francis Rumsey*

Digital Sound Processing for Music and Multimedia
Ross Kirk and Andy Hunt

The Digital Audio CD and Resource Pack *Mark Erne*

Computer Sound Synthesis for the Electronic Musician

Eduardo Reck Miranda

Focal Press

OXFORD   AUCKLAND   BOSTON   JOHANNESBURG   MELBOURNE   NEW DELHI

Focal Press
An imprint of Butterworth-Heinemann
Linacre House, Jordan Hill, Oxford OX2 8DP
225 Wildwood Avenue, Woburn, MA 01801-2041
A division of Reed Educational and Professional Publishing Ltd

ℛ   A member of the Reed Elsevier plc group

First published 1998
Reprinted 2000

British Library Cataloguing in Publication Data
A catalogue record for this book is available from the British Library

Library of Congress Cataloguing in Publication Data
A catalogue record for this book is available from the Library of Congress

ISBN 0 240 51517 X

Composition by Scribe Design, Gillingham, Kent
Printed and bound in Great Britain

FOR EVERY VOLUME THAT WE PUBLISH, BUTTERWORTH-HEINEMANN
WILL PAY FOR BTCV TO PLANT AND CARE FOR A TREE.

Contents

Preface

Computer Sound Synthesis for the Electronic Musician is intended to introduce computer sound synthesis techniques and synthesis programming to students, researchers, musicians and enthusiasts in the field of music technology. In order to be able to make optimum use of this book, the reader should have a basic understanding of acoustics and should be familiar with some fundamental concepts of music and technology. Also, a minimum degree of computer literacy is desirable in order to take full advantage of the systems provided on the accompanying CD-ROM.

The art of sound synthesis is as important for the electronic musician as the art of orchestration is for composers of symphonic music. Both arts deal with creating sonorities for musical composition. The main difference between them is that the former may also involve the creation of the instruments themselves. Those who desire to make their own virtual orchestra of electronic instruments and produce new original sounds will find this book very useful. It examines the functioning of a variety of synthesis techniques and illustrates how to turn a personal computer into a powerful and flexible sound synthesiser. The book also discusses a number of ongoing developments that may play an important role in the future of electronic music making, including the use of artificial intelligence techniques in synthesis software and parallel computers.

The accompanying CD-ROM contains examples, complementary tutorials and a number of synthesis systems for PC and

Macintosh platforms, ranging from low-level synthesis programming languages to graphic front-ends for instrument and sound design. These include fully working packages, demonstration versions of commercial software and experimental programs from major research centres in Europe and North and South America.

Computer sound synthesis is becoming increasingly attractive to a wide range of musicians. On the one hand, and with very few exceptions, manufacturers of mainstream MIDI-based synthesisers have not yet been able to produce many known powerful synthesis techniques on an industrial scale. Moreover, the MIDI communication protocol itself is not fully capable of providing flexible and expressive parameters to control complex synthesis techniques. On the other hand, the sound processing power of the personal computer is increasing, and is becoming more affordable. Computers are highly programmable and any personal computer will soon be able to run software in real time, capable of synthesising sounds using any technique that one could imagine.

Musicians often may not wish to use preset timbres but would prefer to create their own instruments. There are, however, a number of ways to program instruments on a computer and the choice of a suitable synthesis technique is crucial for effective results. Some techniques may perform better than others for some specific timbres, but there are no definite criteria for selection; it is basically a matter of experience and taste. In general, the more intuitive and flexible the technique, the more attractive it tends to be. For example, those techniques whose parameters provide meaningful ways to design instruments and timbres are usually preferred to those whose parameters are based entirely upon abstract mathematical formulae.

A growing number of synthesis techniques have been invented and used worldwide. There is not, however, a generally agreed taxonomy to study them. Indeed, the synthesiser industry sometimes makes the situation worse by creating various marketing-oriented labels for what might essentially be the same synthesis paradigm. For the purpose of this book, the author has devised a taxonomy, assuming that synthesis techniques work based upon a *model*. For instance, some synthesis models tend to employ loose mathematical abstractions, whereas others attempt to mimic mechanical-acoustic phenomena. Synthesis techniques may thus be classified into four classes according to their modelling approach: *loose modelling*, *physical modelling*, *time modelling* and *spectral modelling*. It is important to observe that the boundaries of this taxonomy may overlap; some techniques may qualify for more than one class. In these cases, the author has deliberately highlighted those qualities that best fit his suggested classification.

Loose modelling techniques (discussed in Chapter 3) tend to provide synthesis parameters that bear little relation to the acoustic world. They are usually based entirely upon conceptual mathematical formulae. It is often difficult to predict the outcome and to explore the potential of a loose model. Frequency modulation (FM) is a typical example of loose modelling. FM is a powerful technique and extremely easy to program but difficult to operate; computer simulations of Yamaha's acclaimed DX7 synthesiser, for example, are relatively easy to implement on a computer. Nevertheless, it is apparent that the relationship between a timbre and its respective synthesis parameters is far from intuitive. Apart from John Chowning, the inventor of FM synthesis, only a few people have managed to master the operation of this technique.

Physical modelling and spectral modelling attempt to alleviate this problem by providing less obscure synthesis parameters; both support the incorporation of natural acoustic phenomena. The fundamental difference between physical and spectral modelling techniques is that the former tends to model a sound at its source, whilst the latter tends to model a sound at the basilar membrane of the human ear.

In general, physical modelling techniques work by emulating the functioning of acoustic musical instruments (see Chapter 4). The key issue of physical modelling is the emulation of acoustic sound generators rather than of the sounds themselves. For example, whilst some synthesis techniques (e.g. additive synthesis) attempt to produce a flute-like sound using methods that have little resemblance to the functioning of the flute, a physical modelling technique would attempt to synthesise it by emulating a jet of air passing through a mouthpiece into a resonating pipe.

The implementation of a physical model is not straightforward. However, once the model is implemented, it is not complicated to interpret the role of their synthesis parameters. Take, for example, a singing voice-like instrument. A loose model using FM, for instance, would provide relatively complex synthesis parameters, such as modulation index and frequency ratio. Conversely, a physical model using waveguide filtering, for instance, would provide more easily interpreted synthesis parameters, such as air pressure, vocal tract shape and throat radiation output.

Spectral modelling techniques have their origins in Fourier's Theorem and the additive synthesis technique (see Chapter 6). Fourier's Theorem states that any periodic waveform can be modelled as a sum of partials at various amplitude envelopes and time-varying frequencies. Additive synthesis is accepted as perhaps the most powerful and flexible spectral modelling method, but it is difficult and expensive to run. Musical timbres are composed of dozens of time-varying partials, including harmonic, non-harmonic and noise components. It would require dozens of oscillators, noise generators and envelopes to simulate musical timbres using the classic additive technique. The specification and control of the parameter values for these components are difficult and time consuming. Alternative methods have been proposed to improve this situation, by providing tools to produce automatically the synthesis parameters from the analysis of sampled sounds. Note that the analysis techniques used here store filter coefficients rather than samples. The great advantage of spectral modelling over plain sampling is that musicians can manipulate these coefficients in a variety of ways in order to create new sounds. Sound morphing, for example, can be achieved by varying the coefficients accordingly.

Finally, time modelling techniques approach synthesis from a time domain perspective. The parameters of time modelling synthesis tend to describe sound evolution and transformation of time-related features; e.g. in terms of time lapses. Examples of time modelling techniques include granular synthesis and sequential waveform composition; time modelling techniques are discussed in Chapter 5.

It is generally agreed that no single synthesis technique will ever be able to fully satisfy the exacting musician. In the not too distant future, musicians will probably prefer to give up collecting synthesiser modules and keyboards, and opt for a personal computer running powerful sound synthesis software.

I would like to express my gratitude to all contributors who have kindly provided the materials for the CD-ROM: Richard Moore (University of California, San Diego), Robert Thompson (Georgia State University), Bill Schottstaedt and Nicky Hind (University of Stanford), Max Malcolm (The Karnataka Group), Kenny McAlpine (University of Glasgow), Stephan Schmitt (Native Instruments), Nicolas Fournel (Synoptic), Mike Berry (Mills College), Greg Lee (Seer Systems), Aluizio Arcela (University of Brasília), Xavier Rodet and Adrien Lefevre (Ircam), Trevor Wishart, Archer Endrich and Richard Dobson (CDP), Jacques

Chareyron and Raffaele de Tintis (University of Milan), Arun Chandra (University of Illinois), Xavier Serra (Pompeu Fabra University), Joe Wright (Sonic Design), Joerg Stelkens (büro </Stelkens>) and Rasmus Enkman (Stockholm Electroacoustic Music Studios).

I would also like to thank Francis Rumsey, Robert Thompson and Dennis Miller who read the manuscript draft and made invaluable comments and suggestions.

Finally, I am indebted to my wife Alexandra for her extraordinary encouragement and support at all times. This book is dedicated to my parents, Luiz Carlos and Maria de Lurdes, who encouraged me to be a musician.

Eduardo Reck Miranda

1 Computer sound synthesis fundamentals

1.1 Digital representation of sound

Sound results from the mechanical disturbance of some object in a physical medium such as air. These mechanical disturbances generate vibrations that can be represented as electrical signals by means of devices (for example, a microphone), that convert these vibrations into time-varying voltage. The result of the conversion is called an *analog signal*. Analog signals are continuous in the sense that they consist of a continuum of values, as opposed to stepwise values.

An analog signal can be recorded onto a magnetic tape using electromagnetic technology. In order to play back sounds recorded on magnetic tape, the signal is scanned and sent to a loudspeaker that reproduces the sound vibrations in the air. Analog synthesisers basically function by creating sounds from scratch, using electronic devices capable of producing suitable signals to vibrate the loudspeakers.

Computers, however, are digital rather than analog machines. In other words, computers work based upon discrete mathematics. The adjective 'discrete' is defined in *The Collins Concise Dictionary of the English Language* as 'separate or distinct; consisting of distinct or separate parts' as opposed to continuous phenomena. Discrete mathematics is employed when entities are counted rather than weighed or measured; for example, it is suitable for tasks involving the relations between one set of objects and

1

another set. In this case, calculations must involve only finite and exact numbers. In order to illustrate the difference between the analog and the digital domains, let us compare an analog clock with a digital clock. The analog clock does not give us a precise time value. The hands of the clock move continuously and by the time we read them, the clock has already moved forwards. Conversely, the digital clock gives a sense of precision because the digits of its display do not move continuously; they jump from one time lapse to another. Whilst the analog clock is seen to *measure* time, the digital clock is seen to *count* it.

The main difficulty in using the computer for sound synthesis is that it works only with *discrete domains* and the knowledge of sounds that science has generated throughout history is essentially analog. Moreover, computers can deal only with *binary numbers*. In contrast to the decimal numeric system, which uses ten different symbols (i.e. from 0 to 9) to represent numbers, the binary system uses only two symbols: 0 and 1. Computers are made of tiny electronic switches, each of which can be in only one of two states at a time: on or off, represented by the digits '1' and '0', respectively. Consequently, the smallest unit of information that the computer can handle is the *bit*, a contraction of the term *binary digit*. For instance, the decimal numbers 0, 1, 2 and 3 are represented in the binary system as 0, 1, 10 and 11 respectively.

Computers are normally configured to function based upon strings of bits of fixed size, called *words*. For example, a computer configured for 4-bit words would represent the decimal numbers 0, 1, 2 and 3 as 0000, 0001, 0010 and 0011. Note that the maximum number that four bits can represent is 1111, which is equivalent to 15 in the decimal system. In this case, a 4-bit computer seems to be extremely limited, but in fact, even a much larger word size would present this type of limitation here. Indeed, most currently available computers use 16-, 32- or 64-bit words, but they represent numbers in a slightly different way. Computers actually consider each digit of a decimal number individually. Hence our example 4-bit computer would use two separate words to represent the decimal number 15, one for the digit 1 and another for the digit 5: 15 = 0001 0101.

The binary representation discussed above is extremely useful because computers also need to represent symbols other than numbers. Every keyboard stroke typed on a computer, be it a number, a letter of the alphabet or a command, needs to be converted into a binary number-based representation before the computer can process it. Conversely, output devices, such as the video display and the printer, need to convert the binary-coded information to suitable symbols for rendition. Manufacturers assign arbitrary codes for symbols other than numbers: for instance, the letter A = 10000001 and the letter B = 10000010. Whilst part of this codification is standard for most machines (e.g. the ASCII code), a significant proportion is not, which leads to one of the causes of incompatibility between different makes of computers.

An analog sound signal is a continuous one that is incompatible with the way in which the computer represents data. In order to process sounds on the computer, the analog signal must be converted into a digital format; that is, the sound must be represented using binary numbers. Conversely, the digital signal must be converted into analog voltage in order to play a sound from the computer. The musical computer must therefore be provided with two types of data converters: analog-to-digital (ADC) and digital-to-analog (DAC).

The conversion is based upon the concept of *sampling*. The sampling process functions by measuring the voltage (that is, the amplitude) of the continuous signal at intervals of equal duration. Each measurement value is called a *sample* and is recorded in binary format. The result of the whole sampling process is therefore a sequence of binary numbers corresponding to the voltages at successive time lapses (Figure 1.1).

Figure 1.1 The sampling process functions by measuring the amplitude of a continuous signal at intervals of equal duration. Note that the converter rounds off the measurements in order to fit the numbers that it can process

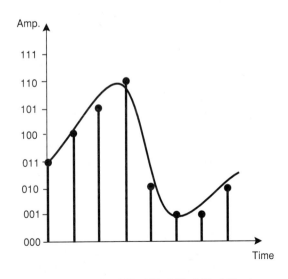

000 100 101 110 010 001 001 010 etc.

Audio samples can be stored on any digital medium, such as tape, disk or computer memory, using any recording technology available, including electromagnetic and optic technology; for example, Winchester disk drive (or hard disk), Digital Audio Tape (DAT) or Compact Disc (CD). In order to synthesise sounds from scratch, the computer must be programmed to generate the right streams of samples. The advantage of digital sound representation over analog representation is that the former allows for computer manipulation and generation of streams of samples in an infinite variety of ways. Furthermore, the possibilities of digital sound synthesis are much greater than analog synthesis.

1.1.1 The sampling theorem

The number of times the signal is sampled in each second is called the *sampling rate* (or sampling frequency) and it is measured in Herz (Hz).

The sampling theorem states that in order accurately to represent a sound digitally, the sampling rate must be higher than at least twice the value of the highest frequency contained in the signal. The faster the sampling rate, the higher the frequency that can be represented, but the greater the demands for computer memory and power. The upper limit of human hearing is approximately 18 kHz, which implies a minimum sampling rate of 36 kHz (i.e. 36 000 samples per second) for high fidelity. The sampling rate frequently used in computer sound synthesis systems is 44 100 Hz.

1.1.2 Quantisation noise

The amplitude of a digital signal is represented according to the scale of a limited range of different values. This range is determined by the resolution of the ADC and DAC. The resolution of the converters depends upon the size of the word used to represent each sample. For instance, whereas a system with 4-bit resolution would have only 16 different values (2^4) to measure the amplitudes of the samples, a system with 16-bit resolution would have 65 536 different values (2^{16}). The higher the resolution of the converters, the better the quality of the digitised sound. The sampling process normally rounds off the measurements in order to fit the numbers that the converters can deal with (Figure 1.2). Unsatisfactory lower resolutions are prone to cause a damaging loss of sound quality, referred to as the *quantisation noise*.

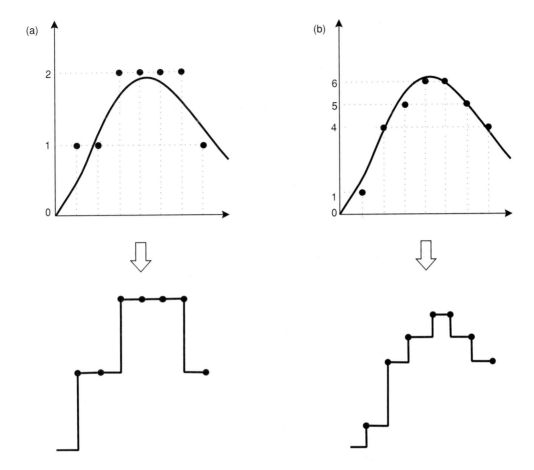

Figure 1.2 The higher the resolution of the converters, the better the accuracy of the digitised signal

1.1.3 Nyquist frequency and aliasing distortion

Nyquist frequency is the name of the highest frequency that can theoretically be represented in a digital audio system. It is calculated as half of the value of the sampling rate. For instance, if the sampling rate of the system is equal to 44 100 Hz, then the Nyquist frequency is equal to 22 050 Hz.

The ADC needs at least two samples per waveform cycle in order to represent the frequency of the sound; if not, then the frequency information is lost. Digital recording systems place a low-pass filter before the ADC in order to ensure that only signals below the Nyquist frequency enter the converter. Otherwise, the conversion process creates foldback frequencies, thus causing a phenomenon known as *aliasing distortion* (Figure 1.3).

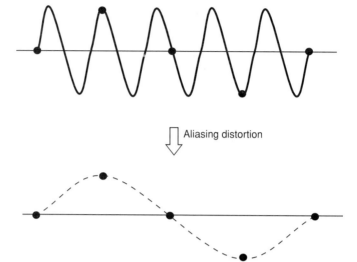

Figure 1.3 An analog–to–digital converter needs at least two samples per cycle in order to represent a sound. Otherwise the sampling process creates aliasing distortion

Aliasing distortion

1.1.4 Sound file formats

Sound storage

Digital audio may be stored on a computer in a variety of formats. Different systems use different sound file formats, which define how samples and other related information are organised in a computer file. The most basic method to store a sound is to take the stream of samples generated by the synthesis program, or from the ADC, and write them on a file. This method, however, is not flexible because it does not allow for the storage of information other than the raw samples; for example, the sampling rate used, the size of the word or whether the sound is mono or stereo. In order to alleviate this problem, sound files normally include a descriptive data structure, referred to as the sound file header, that encodes the properties of the file. A program that uses the sound file then reads the header and configures the system appropriately. Some sound file headers even allow for the inclusion of text comments and cue pointers in the sound. Frequently used sound file formats are:

- WAVE, adopted by Microsoft and IBM (.wav)
- VOC, adopted by Creative Lab's Sound Blaster (.voc)
- NeXT/Sun, originated by NeXT and Sun computers (.snd)
- AIFF, originated by Apple computers (.aif)
- AVR, adopted by Atari and Apple computers (.avr)

Sound transfer

The convenience of digital storage media is that audio samples can be transferred from one medium to another with no loss of sound quality. But the transfer of audio samples between devices can sometimes be burdensome because of eventual incompatibilities between different equipment.

There are a number of data transfer formats and almost every manufacturer has established their own. From the user's point of view, it is very disappointing to learn that the new piece of equipment you always wanted is finally available for sale but that it is incompatible with the gear available in your studio. Fortunately, the digital audio equipment industry is becoming aware of these inconsistencies and is moving towards more standardised transfer formats. The two most popular transfer formats in use today are AES/EBU and S/PDIF. The AES/EBU format is a serial two-channel format, created by the Audio Engineering Society (AES) and the European Broadcast Union (EBU) for professional equipment. The S/PDIF format (Sony/Philips Digital Interface Format) is similar to the AES/EBU format, but it is intended primarily for domestic equipment and home studio set-ups. For a comprehensive coverage of this subject, refer to Rumsey and Wakinson (1995) and Rumsey (1996).

1.2 Basics of computer programming for sound synthesis

In principle, computers can be programmed to perform or solve almost any imaginable task or problem, with the proviso that the method for resolution can be explicitly defined and the data needed for its realisation can be explicitly represented. In essence, in order to program a computer one needs to write a sequence of instructions specifying how the machine will achieve the required results. This implies that the programmer must know how to resolve the problem or task in order to instruct the machine.

In many ways, a computer program is fairly similar to a recipe for a dish: a recipe gives clear steps to prepare the dish and lists all necessary ingredients. Recipes as well as computer programs must have no ambiguities; e.g. the statement 'cover and simmer for five minutes over low heat' is preferable to 'cover and simmer for some time'. Note, however, that even though the first statement is more precise than the second, the former still implies some background knowledge to interpret it. The term 'low heat', for instance, may carry different meanings to a chef, to a chemist and to an astrophysicist; the chef will probably know better how to ascertain the gas mark of the cooker. People tend to develop specific coding and languages to communicate ideas on a specific subject more efficiently. The BBC Radio 4 Shipping Forecast is a good example of an efficient coding system to clearly communicate the weather forecast to sailors.

A programmer communicates instructions to the computer by means of a programming language. The idiosyncrasies of certain tasks require specific features which may require the design of specific programming languages to deal with them. Many programming languages have been developed for a diversity of purposes and types of computers, for example C, Basic, Fortran, Java, Lisp, Pascal and Prolog, to cite but a few. Thus, whilst Prolog was specially devised for artificial intelligence research, Java was planned for applications involving the Internet. The main advantage of using a language that has been specially designed for specific problems is that its vocabulary normally employs the jargons of the domain application. For example, the command to activate an oscillator in CLM (a programming language for sound synthesis; see Chapter 2) is *oscil*, which is short for oscillator.

For the same reason that numbers, the alphabet and sounds must be converted into binary numbers before the computer can perform any processing, programs also have to be converted into machine code. Machine code is the only language that computers can understand but it is extremely tedious, if not impossible, to write a large complex program using raw binary code; hence the rationale for the so-called high-level programming languages. Such programming languages provide a 'translator' that converts the encoded program into machine code. There are two kinds of translators: *compiler* and *interpreter*. The compiler converts the whole code before executing it. Compilers save the result onto an executable file which can be activated at any time without having to repeat the translation process. Conversely, the interpreter converts each statement of the program and executes it before it proceeds to convert the next statement. Each mode has its own merits and pitfalls but an in-depth discussion of these is beyond the scope of this book.

The first programming language specifically designed for sound synthesis was Music III, created by Max Mathews and Joan Miller at the Bell Telephone Laboratories, USA, in the early 1960s (Manning, 1987). Several subsequent versions soon appeared at Stanford University (Music 10) and Princeton University (Music 4B). Most sound synthesis languages available today are descended in one way or another from these languages, generally referred to as Music N, including the ones provided on the accompanying CD-ROM: pcmusic (developed by F. Richard Moore at the University of California in San Diego, USA), Som-A (developed by Aluizio Arcela at the University of Brasília, Brazil) and CLM (developed by Bill Schottstaedt at the University of Stanford, USA).

1.2.1 Algorithms and control flow

An algorithm is a sequence of instructions carried out to perform a task or to solve a problem. They are normally associated with computing, but in fact they abound everywhere, even in cans of shaving foam:

> Begin shaving foam algorithm
> > Shake can well
> > Wet face
> > Hold can upright
> > Release foam onto fingers
> > Smooth onto face
> > Shave
> > Rinse face
> End shaving foam algorithm

Software engineers distinguish between an algorithm and a program. An algorithm is an abstract idea, a schematic solution or method, which does not necessarily depend upon a programming language. A program is an algorithm expressed in a form suitable for execution by a computer. The program is the practical realisation of the algorithm for use with a computer.

Depending on the complexity of the task or problem, a program may require a myriad of tangled, complex algorithms. In order to aid the design of neat and concise algorithms, the software engineering community has developed a variety of programming schemes and abstract constructs, many of which are now embedded in most languages, including those used for sound synthesis. The more widespread of these programming schemes include *encapsulated subroutines*, *path selection* and *iteration*.

Encapsulated subroutines

One of the most fundamental practices of computer programming consists of organising the program into a collection of smaller subprograms, generally referred to as *subroutines* (also known as *macro-modules, procedures* or *functions*). In this way, the same subroutine can be used more than once in a program without the need for rewriting. Subroutines may be stored in different files, frequently referred to as *libraries*; in this case, the compiler or interpreter merges the subroutines invoked by the program.

Most programming languages provide a library with a large number of subroutines which significantly facilitate the ease of the programming task. Programmers are actively encouraged to create their own library of functions as well. On highly developed programming desktops, the creation of a new program may simply require the specification of a list of subroutine calls.

In the example below, *algorithm A* is composed of a sequence of four instructions:

> Begin algorithm A
>   Instruction 1
>   Instruction 2
>   Instruction 3
>   Instruction 4
> End algorithm A

Suppose that this algorithm performs a task that will be requested several times by various sections of a larger program. Instead of writing the whole sequence of instructions again, the algorithm could be encapsulated into a subroutine called *Procedure A*, for example. This subroutine can now be invoked by other algorithms as many times as necessary. For example:

> Begin algorithm B
>   Instruction 5
>   Procedure A
>   Instruction 6
> End algorithm B

The method or command for encapsulating a subroutine may vary considerably from one language to another, but the idea is essentially the same in each case. As far as sound synthesis is concerned, synthesis languages provide a library of ready-made subroutines which are the building blocks used to assemble synthesis instruments.

Path selection

The instructions and subroutines must be performed in sequence and in the same order as they were specified. There are cases, however, in which an algorithm may have to select an execution path from a number of options. The most basic construct for path selection is the *if-then* construct. The following example illustrates the functioning of the *if-then* construct. Note how *Instruction 2* is executed only if 'something' is true:

```
Begin algorithm A
    Instruction 1
    IF something
        THEN Instruction 2
    Instruction 3
    Instruction 4
End algorithm A
```

Another construct for path selection is the *if-then-else* construct. This is used when the algorithm must select one of two different paths. For example:

```
Begin algorithm B
    Instruction 1
    IF something
        THEN Instruction 2
        ELSE   Instruction 3
                Instruction 4
    Instruction 5
End algorithm B
```

In this case, if 'something' is true, then the algorithm executes *Instruction 2* and immediately jumps to execute *Instruction 5*. If 'something' is not true, then the algorithm skips *Instruction 2* and executes *Instruction 3* and *Instruction 4*, followed by *Instruction 5*.

The *if-then* and the *if-then-else* constructs are also useful in situations where the algorithm needs to select one of a number of alternatives. For example:

```
Begin algorithm C
    Instruction 1
    IF something
        THEN Instruction 2
    IF another thing
        THEN Instruction 3
        ELSE Instruction 4
    IF some condition
        THEN Instruction 5
    Instruction 6
End algorithm C
```

Finally, there is the *case* construct. This construct is used when the selection of a path depends upon the multiple possibilities of one single test. In the example below, there are three 'possibilities' (X, Y and Z) that satisfy 'something'. Each 'possibility' triggers the execution of a different *Instruction*:

```
Begin algorithm D
    Instruction 1
    CASE something
        possibility X perform Instruction 2
        possibility Y perform Instruction 3
        possibility Z perform Instruction 4
    Instruction 5
End algorithm D
```

In this example, *algorithm D* executes either *Instruction 2*, *Instruction 3* or *Instruction 4* depending on the assessment of 'something'; only one of these three instructions will be executed.

Iteration

Iteration (also referred to as loop) allows for the repetition of a section of the program a number of times. There are many ways to set up an iteration, but all of them will invariably need the specification of either or both the amount of repetitions or a condition to terminate the iteration. The most common constructs for iteration are *do-until*, *while-do* and *for-to-do*.

The following example illustrates the *do-until* construct:

```
Begin algorithm A
    Instruction 1
    DO  Instruction 2
        Instruction 3
        Instruction 4
    UNTIL something
    Instruction 5
End algorithm A
```

In this case, the algorithm executes *Instruction 1* and then repeats the sequence *Instruction 2*, *Instruction 3* and *Instruction 4* until 'something' is true. A variation of the above algorithm could be specified using the *while-do* construct, as follows:

```
Begin algorithm B
    Instruction 1
    WHILE something
    DO   Instruction 2
         Instruction 3
         Instruction 4
    Instruction 5
End algorithm B
```

The later algorithm (*algorithm B*) also repeats the sequence *Instruction 2*, *Instruction 3* and *Instruction 4*, but it behaves slightly different from *algorithm A*. Whereas *algorithm A* will execute the sequence of instructions at least once, indifferent to the status of 'something', *algorithm B* will execute the sequence only if 'something' is true right at the start of the iteration.

The *for-to-do* construct works similarly to the *while-do* construct. Both may present some minor differences depending on the programming language on hand. In general, the *for-to-do* construct is used when an initial state steps towards a different state. The repetition continues until the new state is eventually reached. For example:

```
Begin algorithm C
    Instruction 1
    FOR initial state TO another state
    DO   Instruction 2
         Instruction 3
         Instruction 4
    Instruction 5
End algorithm A
```

In this case, the sequence *Instruction 2*, *Instruction 3* and *Instruction 4* will repeat until the state reaches 'another state'.

1.2.2 Unit generators

Synthesis languages generally provide a number of synthesis subroutines, usually referred to as *unit generators* and in this case, synthesis instruments are programmed by interconnecting a number of unit generators. Originally, the design of these unit generators was based upon the principle that they should simulate the modules of an analog synthesiser. The rationale was that electronic musicians accustomed to the setting up of analog synthesis patches would not find it so difficult to migrate to computer sound synthesis programming. Although computer

Figure 1.4 On a computer, the oscillator works by repeating a template waveform, stored in a lookup table

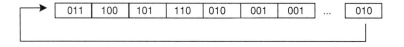

| 011 | 100 | 101 | 110 | 010 | 001 | 001 | ... | 010 |

synthesis technology has evolved from mere analog simulation to much more diverse and sophisticated trains of thought, a great deal of the original unit generator ideology still applies for various currently available synthesis programming languages and systems.

The most basic, but by no means the least important, unit generator of a synthesis programming language or system is the *oscillator*. The concept of a computer-based oscillator differs from the analog synthesiser's oscillator. Whereas the analog oscillator produces only a sinusoidal waveform, a computer-based oscillator can produce any waveform, including, of course, the sinusoidal one. On a computer, the oscillator works by repeating a template waveform, stored on a *lookup table* (Figure 1.4); note here the appearance of the iteration programming scheme introduced earlier. The speed at which the lookup table is scanned defines the frequency of the sound. The lookup table contains a list of samples for one cycle of a waveform, which does not necessarily need to be a sinusoid. For this reason, the specification of an oscillator on a computer always involves at least three parameters: *frequency, amplitude* and *waveform*. The most commonly used symbol for such oscillators in the sound synthesis literature is the elongated semi-circle, with the amplitude and frequency inputs at the top and the signal output at the bottom. The interior of the figure may contain a symbol or a notation defining the nature of the waveform in the lookup table (Figure 1.5).

Function generators form another important class of synthesis units. These generators create the lookup tables for the oscillators and tables for sound transformation and control. Function generators fill data tables with values produced according to specific procedures or mathematical formulae, such as trigonometric functions and polynomials. The length of the function table is generally specified as a power of two, such as 512 or 1024 samples.

Amp.   Freq.

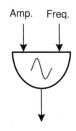

Figure 1.5 The most commonly used symbol for an oscillator

A third class of synthesis units comprises *signal modifiers* such as low- and high-pass filters (see Chapter 4).

The following example illustrates the codification in pcmusic of the instrument in Figure 1.6; refer to Chapter 2 for an introduction to pcmusic:

```
#include <cmusic.h>
#include <waves.h>
instrument 0 simple;
    osc b2 p6 p5 f2 d;
    osc b1 b2 p7 f1 d;
    out b1;
end;
SINE(f1);
ENV(f2);
```

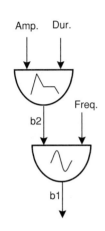

Amp.   Dur.

Freq.

b2

b1

Figure 1.6 A simple instrument

Note that the amplitude of the oscillator is controlled by an envelope, which in this case is implemented by means of another oscillator unit. The lookup table for the 'envelope oscillator' holds the shape of the envelope and is read only once per note. In this case, the frequency of the 'envelope oscillator' must be set to the inverse of the duration of the note (*freq = 1 / duration*).

Envelopes are very useful for adding time-varying control to synthesis instruments and the example in Figure 1.6 illustrates only one of the many possible ways to implement them. Synthesis languages usually provide a few specially built envelope generators (e.g. *seg* in pcmusic and *make-env* in CLM).

1.2.3 Playing the instruments

Once an instrument has been specified, then the computer needs information to play it. The method will vary from system to system. The standard way is to write a score for the computer, as if one were to write a piano roll for a mechanical pianola. This is the case for Music N types of programming languages such as pcmusic, Som-A and CLM. Some programming systems allow for the use of controllers, such as a MIDI keyboard, attached to the computer, so that it can be played as a performance instrument. Examples of such systems are: Audio Architect, GENERATOR and GrainWave (see Chapter 2).

On certain Music N type software, the musician normally creates two separate files, one containing the instrument, or a set of instruments, called an *orchestra file*, and another the parameters to 'play' the instruments, called a *score file*. The synthesis program (i.e. the compiler or the interpreter) reads both files, converts them into machine-level instructions, builds the instruments of the 'orchestra' and feeds them the parameters from the score file. The final audio signal is either written onto a sound file or is played back during processing, or both. In languages such as pcmusic, however, the instruments and the score are normally placed on one single file.

The main drawback of the Music N type languages is that they tend to require explicit manual specification of scores. For example, if an instrument requires 50 parameters to play a single note, then it would require the specification of 5000 parameter values to play 100 notes. More progressive languages alleviate this problem by providing the ability to embed formulae or full programming facilities within the score (e.g. CLM). Also, there are a few programs for synthesis control and score processing, specially designed to aid the composition of larger sequences of sound events.

Those systems which allow for MIDI-based control have the advantage that they can be used in conjunction with MIDI sequencing software to produce the scores. However, the limitations of the MIDI protocol itself may render this facility inappropriate for more elaborate control of synthesis instruments.

2 Introduction to the software on the accompanying CD-ROM*

On the accompanying CD-ROM there are a number of synthesis softwares of various styles, ranging from generic programming languages to programs dedicated to specific synthesis techniques. They will be used to illustrate the techniques presented in the following chapters of this book but notice, however, that the choice of a system to illustrate an example does not necessarily mean that other systems cannot do the same job; it is just a matter of personal preference. Of course, the more generic the software, the greater the number of synthesis techniques it can support.

The remaining sections of this chapter give an overview of these softwares, but the reader is strongly advised to refer to the documentation on the CD-ROM for more in-depth information on each of them. The documentation for most of the programs includes comprehensive tutorials in addition to the user's guide.

For didactic purposes, the softwares have been classified into four groups according to their style of operation:

- Synthesis programming languages
- Synthesis programming systems
- Software synthesisers
- Non-generic synthesis software

*For detailed CD-ROM instructions please see page 197.

Whereas synthesis programming languages incorporate programming languages in which synthesis algorithms are typed in textual form according to a prescribed syntax, synthesis programming systems include systems in which synthesis algorithms are implemented by means of some sort of graphic interface or visual aid. Software synthesisers also resort to graphic interfaces but their programming abilities are severely constrained by predefined synthesis arrangements. Synthesis programming in these cases is normally associated with the setting up of a number of parameters for a super 'smart alec' machine. Non-generic synthesis software, as the name suggests, are systems that were built to perform a specific synthesis task or technique.

A number of the softwares on the CD-ROM are fully working versions, but those of a commercial nature may have been abridged according to the policy of each manufacturer. As far as this book is concerned, any should be able to run at least the examples cited here.

2.1 Synthesis programming languages

2.1.1 pcmusic

The synthesis programming language pcmusic is the DOS version of cmusic, for PC-compatible computers which was originally designed to run on Unix machines. The cmusic language is a classic Music N type of synthesis programming language devised in 1980 by F. Richard Moore at the University of California, San Diego, USA (Moore, 1990).

The system is provided with a large library of unit generators and signal processing subroutines that have evolved over the years. It works as a compiler, therefore it was not designed to work in real time. There are virtually no limitations to the complexity of the instruments in pcmusic because the system does not bother with processing speed. Slower processors naturally take longer to compile, but faster ones do not necessarily improve the instrument design capabilities of the language.

In classic Music N types of programming languages such as pcmusic the electronic musician designs instruments by writing instructions specifying interconnections of unit generators and setting up the lookup tables containing the waveforms for oscillators, envelopes and other controllers. Once the instruments are specified, then the musician must put together a list of notes to play them. Each note of the list consists of start time and duration specifications, followed by a number of synthesis parameter values, as required by the instrument that will play it. The term 'note' here refers to a generalised sound event of any complexity. Both instruments and notes are normally saved onto one single file, called *the score*.

The values for global synthesis parameters, such as sampling rate and number of channels, must be specified in a file called *Pcmusic.ini* normally found in the current working directory. This file contains pcmusic specific information about the configuration of the system. Novices are therefore advised not to make major changes in the *Pcmusic.init* file unless they are completely sure of what they are doing.

A pcmusic score is typically divided three major sections:

- Instrument specifications
- Lookup table definitions
- Note list

The specification of instruments

In this section one defines the architecture of the instruments by specifying a certain configuration of unit generators. The general form of an instrument is as follows:

instrument \<start time\> \<name\> ;
\<unit generator instruction 1\> ;
\<unit generator instruction 2\> ;
...
\<unit generator instruction n\> ;
end ;

Every instrument begins with the instruction *instrument* followed by a starting time (normally set to zero) and a label, and closes with the instruction *end*. The body of the instrument contains unit generator instructions. Each instruction line begins with the name of the respective unit generator (e.g. *osc* for an oscillator; *out*, for audio output; *flt* for a filter, etc.) and it is followed by a list of parameters. The following example shows the codification of a simple instrument:

instrument 0 simple ;
   osc b1 p5 p6 f1 d ;
   out b1 ;
end ;

The instrument, labelled as *simple*, has an oscillator (*osc*) which outputs samples to the audio output unit (*out*) via a variable called *b1*. The amplitude and frequency values for the oscillator will be specified in the note list via two variables, *p5*, and *p6*, respectively. The waveform for this oscillator is referred to as *f1* and it should be specified elsewhere in the score. The last parameter of the oscillator is related to its initial phase and the 'value' *d* is usually employed. It is not necessary to concentrate on this parameter at this stage; more details about it can be found in Robert Thompson's tutorial on the accompanying CD-ROM.

The syntax of pcmusic is relatively simple. Each instruction line of a program always begins with a pcmusic command and closes with a semicolon (;). A number of parameter fields are arranged between the initial command and the final semicolon. Each parameter field may be a single value, a variable or an expression, which are separated by blanks or commas. Note that a parameter field itself cannot contain blank spaces.

The specification of lookup tables

The pcmusic package has a number of lookup table generators referred to as *gen* followed by a number; e.g. *gen1* (generates a straight line), *gen4* (produces exponential curves), *gen5* (generates sinusoidal partials), to cite but a few. The instruction *generate* is used to activate a certain *gen* subroutine, as follows:

generate <time> <gen subroutine> <table name> <list of parameters> ;

Two examples using *gen5* to produce sinusoidal components are given as follows:

```
generate 0 gen5 f1 1,1,0 ;
generate 0 gen5 f2 1,1,0 5,1/2,0 ;
```

The generator *gen5* produces a combination of sinusoidal partials described by triplets of parameters specifying the harmonic number, the relative amplitude of the harmonic and its phase offset, respectively. The first example generates a lookup table called as *f1* containing a single sinewave; i.e. harmonic number = 1, relative amplitude = 1 and phase offset = 0. The second example generates table *f2* containing a waveform resulting from the addition of two sinewaves: a fundamental and its fifth harmonic with halved amplitude.

The specification of note lists

The format of the specification of note lists is as follows:

note <start time> <instrument> <duration> <synthesis
parameters>;
note <start time> <instrument> <duration> <synthesis
parameters>;
note <start time> <instrument> <duration> <synthesis
parameters>;
etc. ...

The following example illustrates a sequence of four notes for
the instrument *simple* above. Note the use of explicit symbols for
units such as dB and Hz for amplitude and frequency values,
respectively. Start time and duration values are specified in
seconds.

```
note 0 simple 1 -6 dB 370.0 Hz ;
note 1 simple 1 -6 dB 425.3 Hz ;
note 2 simple 2 -3 dB 466.2 Hz ;
note 4 simple 2 0 dB 622.3 Hz ;
```

A complete example

A complete pcmusic score can now be put together using the
examples given above. The *terminate* instruction is used to
indicate the end of the score. Pcmusic allows for insertion of
comments within brackets in the code:

```
{-----------------------------------
This is an example of a simple
instrument in pcmusic.
-----------------------------------------
Initialisation:
-----------------------------------------}
#include <cmusic.h>
#include <waves.h>
{-----------------------------------------
Instrument specification:
-----------------------------------------}
instrument 0 simple ;
    osc b1 p5 p6 f1 d ;
    out b1 ;
    end ;
{-----------------------------------------
Lookup tables: f1 is a sinewave
-----------------------------------------}
generate 0 gen5 f1 1,1,0 ;
{-----------------------------------------
```

Note list:

```
----------------------------------------}
note 0 simple 1 -6 dB 370.0 Hz ;
note 1 simple 1 -6 dB 425.3 Hz ;
note 2 simple 2 -3 dB 466.2 Hz ;
note 4 simple 2 0 dB 622.3 Hz ;
terminate ;
```

One of the most interesting features of pcmusic is that it has a pre-processor built into its compiler. When the compiler is activated, the score is pre-processed before the compilation takes place. Lines beginning with *#include* instruct the pre-processor to include a copy of a file at that point of the code. In the example above, the pre-processor will include the files *Cmusic.h* and *Waves.h*. These files contain the code for a variety of subroutines that the system needs for the compilation of the instrument; e.g. the code for the unit generators.

The pre-processor also watches for *#define* instructions, which are used to create symbolic labels for synthesis parameters. For example, the statements

```
#define AMP p5
#define FREQ p6
```

instruct the compiler that the symbols 'AMP' and 'FREQ' can be used anywhere in the score in place of *p5* and *p6*, respectively:

```
osc b1 AMP FREQ f1 d ;
```

Finally, in addition to being able to synthesise sounds from scratch using oscillators and mathematically created lookup tables, pcmusic also has the ability to incorporate recorded sounds for manipulation. This leads to a wide range of synthesis possibilities not otherwise available on some synthesis systems.

On the CD-ROM the reader will find a very comprehensive tutorial and a wide variety of examples specially prepared for this book by the composer Robert Scott Thompson of Georgia State University in Atlanta, Georgia, USA.

2.1.2 CLM

CLM, created by Bill Schottstaedt at Stanford University's Music Department, USA, stands for Common Lisp Music (Schottstaedt, 1994). CLM is not an autonomous programming language specifically designed for sound synthesis but a library of unit generators and other miscellaneous synthesis tools (called *functions*, in Lisp jargon) for the Common Lisp programming language. The great advantage of CLM is that the electronic musician has the full potential of Lisp, which is a very powerful programming language, in addition to the sound synthesis possibilities of a Music N type synthesis software.

Currently, CLM runs on NeXT, Macintosh, Silicon Graphics and Sun computers. There is no suitable Common Lisp for Windows to date, but CLM can run on a PC under Linux or Nextstep operational systems (see the user's guide on the CD-ROM).

In order to use CLM the electronic musician must become familiar with the basics of Lisp programming.

Basics of Lisp programming

Lisp presents itself to the user as an interpreter; it works both as a programming language and as an interactive system: it waits for some input or command from the user, executes it, and waits again for further input. Lisp is an acronym for List Processor; almost everything in the Lisp world is a list. In programming terms, a list is a set of elements enclosed between parentheses, separated by spaces. The elements of a list can be numbers, symbols, other lists, and indeed, programs. Examples of lists are:

```
(the number of the beast)
(1 2 3 5 8 13)
(0.666 (21 xyz) abc)
(defun plus (a b) (+ a b))
```

Lisp is excellent at manipulating lists; there are functions that can do almost any operation you can possibly imagine with lists. Programs are lists themselves and they have the same simple syntax of lists. The following example illustrates what happens if one types in a short program for the Lisp interpreter. The sign '<cl>' is the command line prompt indicating that the interpreter is waiting for your command:

```
<cl>(+ 330 336)
666
```

When the interpreter computes (or *evaluates*, in Lisp jargon) a list, it always assumes that the first element of the list is the name of a function and the rest of the elements are the arguments that the function needs for processing. In the above example, the interpreter performs the addition of two numbers; the name of the function is the symbol '+' and the two arguments are the numbers 330 and 336.

If an argument for a function is also a function itself, then the argument is evaluated first, unless otherwise is specified. The general rule is that the innermost function is evaluated first. In the case of various nested functions, the outermost one is normally the last to get evaluated. The Lisp interpreter treats each nested lists or functions independently. For example:

```
<cl>(+ 330 (* 56 (+ 5 1)))
666
```

In this example, the function '+' is evoked with number 330 as the first argument and with the result of the evaluation of (* 56 (+ 5 1)) as the second argument, which in turn evokes the function '*' with number 56 as the first argument and with the result of the evaluation of (+ 5 1) as the second arguments. The result of the innermost step is 6, which is then multiplied by 56. The result of the multiplication is 336 which is finally added to 330. The result of the whole function therefore is 666.

Lisp programs often involve a large number of nested functions and data. In order to keep the code visually tidy, programmers normally break the lines and tabulate the nesting. For example:

```
(+ 330
    (* 56
        (+ 5 1)
    )
)
```

In order to write Lisp functions and programs, the language provides an instruction (or *macro*, in Lisp parlance), named *defun*; short for define function. The following example illustrates the definition of a new function labelled as *plus*:

```
<cl>(defun plus (a b) (+ a b))
PLUS
```

The first argument for *defun* (i.e. second element of the list) is the name of the new function (e.g. *plus*) followed by a list of arguments that the new function should receive when it is called up (e.g. *a* and *b*). The last argument for *defun* is the body of the new function specifying what it should actually do with the arguments. The new *plus* function defined above executes the mere sum of two arguments *a* and *b*, but it could have been a very complex algorithm or an entire program.

A variable is a symbol that represents the memory location for storing information. Variables may be used in Lisp to represent various types of information, such as a number, a list, a program, a single sound sample, a stream of samples or an entire synthesis instrument. The instruction normally used to allocate variables is *setf*. The first argument for *setf* is the label of the variable, and the second argument is the information to be associated to the variable. The following example creates a variable called *series* and associates the list (330 336 666) to this new variable:

```
<cl>(setf series '(330 336 666))
(330 336 666)
```

In Lisp, an inverted comma before a list indicates that its first element is merely a unit of an ordinary list of data, and not a command. From now on, the variable *series* can be used in place of the list (330 336 666). For example, assume that the function *first* outputs the first element of a given list:

```
<cl>(first series)
330
```

Introduction to CLM

A CLM instrument is a Lisp function specifying a particular interconnection of unit generators. The CLM macro for the definition of an instrument is *definstrument*. The general structure of an instrument is:

```
(definstrument name (<param1 param2 param3 ...>)
   (let* ( <setups> )
      (Run
         (loop for i from <first sample> to <last sample>
            do <body of instrument>
            )
         )
      )
   )
```

The structure of the macro *definstrument* is similar to the structure of the macro *defun*. The first argument is a label for the instrument, followed by the second argument which is the list of synthesis parameters that will be used by the instrument. These are the parameters whose values will be provided in the score. The third parameter is the body of the instrument. This has two main sections: *initialisation* and *run-time loop*. The initialisation section normally begins with the *let** statement which is a standard Lisp instruction to declare variables and their initial values. All variables that will be used by the instrument should be declared at this stage. Also, all unit generators that will be used to build the instrument must be declared and properly set up at the initialisation section. The *run-time loop* section begins with the instruction *Run* and it is in this section that the unit generators are combined and interconnected to form the wheels and axles of the instrument.

The following example illustrates the implementation of a simple instrument:

```
;-----------------------------------------
; This instrument produces a sinewave and needs four
; parameters: start time, duration, frequency and amplitude
;-----------------------------------------
(definstrument example (start dur freq amp)
; ------Initialisation section ----------------------
      (let* ((begin (floor (* start sampling-rate)))
            (end (+ beg (floor (* dur sampling-rate))))
            (sine-wave (make-oscil :frequency freq))
            (envel (make-envel    :envelope '(0 0 50 1 100 0)
                                  :scaler amp
                                  :start-time start
                                  :duration dur)))
; -------Run-time loop section -------------------------
            (Run
                (loop for i from begin to end do
                    (outa i (* (env envel) (oscil sine-wave)))
                ) ; close loop
            ) ; close run-time
      ) ; close let
) ; close definstrument
```

Comments in Lisp start with a semi-colon and continue up to the end of the line. The instrument, named as *example*, needs four parameters to produce a sound: start time, duration, frequency and amplitude, labelled as *start*, *dur*, *freq* and *amp*, respectively. There are four instruction lines at the initialisation section. The first two initialise the values for the beginning (*begin*) and ending (*end*) points of the loop subroutine that will produce the samples at the run-time loop section. These two lines rarely need to be changed; people normally copy and paste them from an existing instrument. The other two lines initialise internal structures for the unit generators of the instrument, that is, the *oscillator* and the *envelope*.

Each unit generator of CLM consists of a pair of functions: one of them 'initialises' a data structure for the generator (an operation equivalent to building the lookup tables in pcmusic, for example) and the other runs the generator. The command for initialising the data structures begin with the prefix *make*. For example:

```
(let* (( ...
        (sine-wave (make-oscil : frequency 440))
        ...
```

In this example, the label 'sine-wave' represents an oscillator (*oscil*) which is ready to produce a sinewave of 440 Hz when set in motion via a run-time loop instruction such as:

```
(oscil sine-wave)
```

All CLM unit generators use the same calling sequence such as the one just described for the oscillator; refer to the user's manual on the CD-ROM for the particulars of each unit. CLM uses the same internal table size or cycle length for all units.

Instruments are normally saved on a file and later compiled. Note that, unlike other Music N type software, such as pcmusic, CLM instruments are compiled separately from scores or note lists. The result of the compilation here is a compiled Lisp function and not a sound file. Once the instrument is compiled it can be loaded within the interpreter and be invoked as any other Lisp function. A score, therefore, is just Lisp code invoking instruments.

The instruction to play a instrument is *with-sound*. The following example plays a sequence of four notes on the *example* instrument depicted above:

```
<cl>(with-sound
        (:output "sequence.snd")
        (example 0 1 370.0 0.7)
        (example 1 1 425.3 0.7)
        (example 2 2 466.2 0.8)
        (example 4 2 622.3 1.0))
```

The first argument for *with-sound* specifies global values such as the name of the file used to store the resulting sound. After the global values come the calls for the instrument. Such instructions can, of course, be written using a word processor and saved as a score file to be loaded from the Lisp interpreter.

The great advantage of CLM over most synthesis programming languages is that the score playing the instruments is treated as a program by the Lisp interpreter. That is, the score does not necessarily need to be an explicit list of notes but a program of any complexity. For example:

```
(with-sound
   (:output "bizarre.snd")
   (loop for i from 0 to 24 do
      (let
         ((freq (random 880)))
         (if (> freq 220)
            (example (+ i 1) 1 freq 1)
         ) ; close if
      ) ; close let
   ) ; close loop
) ; close score
```

This program plays a sequence of up to 24 random pitches between 220 Hz and 880 Hz, one second long each.

A complete CLM user's manual (*clm.html*) plus tutorials (*toc.html*) designed by composer Nicky Hind are available on the CD-ROM.

2.2 Synthesis programming systems

2.2.1 Audio Architect

The power and flexibility of synthesis programming languages such as pcmusic and CLM are unquestionably impressive. However, the idea of having to write computer programs to produce sounds may not suit the appetite of some electronic musicians, particularly those used to programming sounds on analog synthesisers. But there again, nothing ventured, nothing gained – some level of computer programming will always be required if one wishes to create non-clichéd, personalised timbres on a computer.

The Karnataka Group, a software company based in London, got in on the ground floor of software synthesis programming and designed Audio Architect: a system that combines computer

Figure 2.1 Audio Architect provides a menu of synthesis modules that mimic the various components of an analog synthesiser

programming with analog-style synthesis. Audio Architect is a synthesis programming system for PC-compatible computers provided with a Microsoft Windows-based graphic interface for building instruments. Like pcmusic and CLM, Audio Architect is also rooted in the concept of unit generators, referred to here as *synthesis modules*. It provides a menu of synthesis modules that mimic the various components of an analog synthesiser, such as oscillators, LFOs, ADSR envelopes, filters and the like (Figure 2.1); see Russ (1996) for a sharp discussion about analog synthesisers.

An instrument, or network in Audio Architect parlance, is simply assembled by dragging synthesis modules from the menu onto a working area (Figure 2.2). The connections between the modules are made simply by placing the cursor over the one module, holding the right mouse button down and dragging the wire to the next module. These connections carry *event* or *audio* signals from one module to another. The main difference between an event signal and an audio signal is that the former is used exclusively to carry control messages and not audio samples; the latter demands far more processing power for computation. Specific parameter settings for each module of the network are specified on a dialogue box which is accessed by double-clicking with the left mouse button over the module (Figure 2.3).

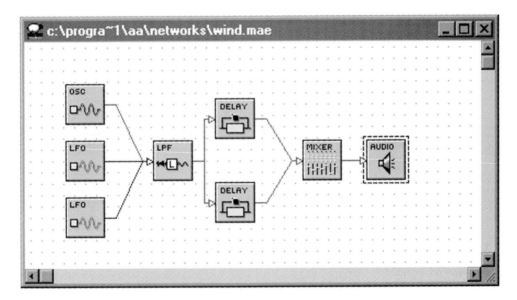

Figure 2.2 Synthesis modules are dragged from the menu onto a working area and the connections between them are made by placing the cursor over one module and dragging the wire to the next module

Figure 2.3 The dialogue box to set the parameters for an oscillator. Specific parameter settings for each module are specified via dialogue boxes which are accessed by double-clicking with the left mouse button over the modules

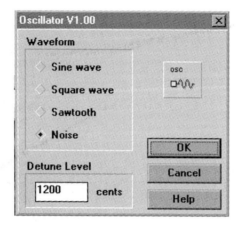

Audio Architect is a thoroughgoing example of an attempt to turn a home computer into a music instrument. Like synthesis programming languages, Audio Architect also generates audio files for later playback. With the proviso that the computer can cope with the complexity of the synthesis instrument at hand, the system offers the possibility of using a MIDI controller to play it in real time as well. Furthermore, the system also reads score files for playing instruments and indeed, Audio Architect features an analog-style sequencer that can step through up to sixteen different synthesis values at a user-specified rate.

On the CD-ROM, Audio Architect is accompanied by a comprehensive tutorial on how to build your own instruments, prepared by Kenny McAlpine at Glasgow University.

2.2.2 GENERATOR

GENERATOR is a synthesis programming system for PC-compatible computers designed by Native Instruments, a German company based in Berlin. The system provides a Windows-based user interface for building instruments which can be played in real time.

Like Audio Architect, GENERATOR instruments are built by combining synthesis modules. GENERATOR, however, provides

Figure 2.4 GENERATOR modules are selected from a pull-down menu and they appear on the workspace as a box with a label and an icon for easy identification. Connections between modules are established by a wire linking the output port of one module to the input port of another

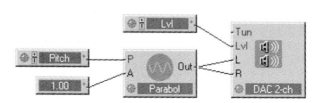

Figure 2.5 The *Properties* window is used to edit the set-up of a module

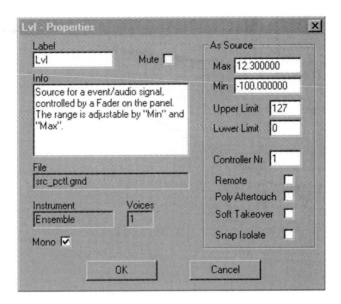

modules for a greater variety of synthesis tasks other than analog-inspired ones, including standard Music N style modules.

The modus operandi of GENERATOR is very similar to Audio Architect. A module is selected by clicking on its respective name on a pull-down menu in the working area. Selected modules appear on the workspace where they can be interconnected to form an instrument (Figure 2.4). Each module is displayed as a box with a label and an icon for easy identification; the label can be edited. Modules normally have input ports on the left-hand side and output ports on the right. Specific characteristics of a module (e.g. the label and a mono/stereo switch) can be edited via the *Properties* window, which can be opened with a double-click on the respective module (Figure 2.5). Connections between modules are established by a wire linking the output port of one module to the input port of another.

One of the most important characteristics of GENERATOR is that it allows for the encapsulation of a configuration of modules into a single 'meta-module', or *macro-module* as it is referred to in the user's manual. A macro-module can be saved in a file and be called up by any other instrument. This facility is very useful for building large complex instruments because it is impracticable to display too many modules on the workspace at once (Figure 2.6). Also, a whole instrument can be saved as a macro-module and be assembled together with other instruments to

Figure 2.6 GENERATOR allows for the encapsulation of a configuration of modules into a single-macro-module. Macro-modules can be stored and called up by another instrument

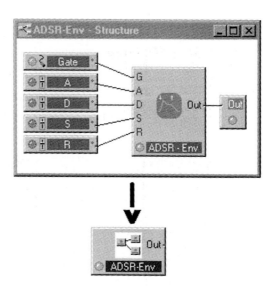

form an *ensemble*. Multiple polyphonic instruments can be assembled and controlled at the same time. The number and complexity of the instruments that can be put together in an ensemble depends on the processing power of the computer.

The system has been engineered primarily to be played in real time. The synthesis parameters can be controlled both by MIDI-controllers and by on-screen 'slide potentiometers', called *faders*. The system has a number of modules for taking in MIDI information from a MIDI source, which can be an external controller or a sequencer running on the same computer. The good news is that the incoming MIDI data can be related to almost any synthesis parameter of an instrument. Alternatively, synthesis parameters can be controlled using the *fader module*. A fader module is displayed on the instrument as a single box containing only its label and the output port. The sliders for displacement with the mouse pointer are displayed on a separate window referred to as the *Panel* (Figure 2.7).

GENERATOR also provides sequencing tools to generate sequences of pre-defined values to assigned synthesis parameters.

2.2.3 *Virtual Waves*

Virtual Waves is a rather sophisticated synthesis programming system for PC-compatible computers which includes not only elementary *sound synthesis* tools but also *sound processing* and *sound analysis* facilities. The system was designed by Nicolas

Figure 2.7 A number of fader modules can be used to control synthesis parameters. The sliders for displacement with the mouse pointer are displayed on a separate window referred to as the *Panel*

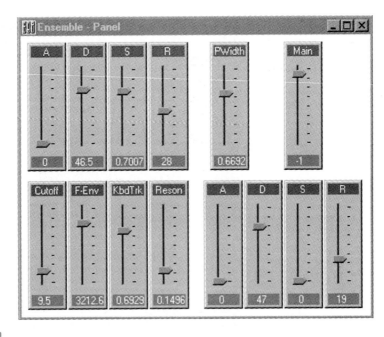

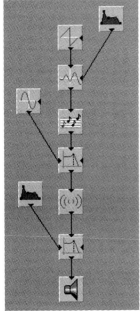

Figure 2.8 The friendly graphic interface of Virtual Waves facilitates the straightforward creation of instruments, which are produced by interconnecting unit generators

Fournel and is commercially available from Synoptic, his own company based in Montreuil, France.

Like Audio Architect and GENERATOR, the user-friendly graphic interface of Virtual Waves greatly facilitates the straightforward creation of instruments (Figure 2.8). Once the synthesis parameters for each module have been adequately adjusted, the sound is then computed. Although Virtual Waves can be very fast for computing simple instruments, the system has not been designed to be 'played' in real time. Also there is no provision for playing back note lists or scores. Virtual Waves should primarily be used as a sound design tool and not as a musical instrument; once the sound has been created, it can either be saved onto a file for later use in some other musical application or transferred to an external sampler or to the RAM of a sound card.

The basic building blocks of Virtual Waves for making instruments are called *modules*. But unlike systems such as GENERATOR, for example, Virtual Waves does not allow for the encapsulation of modules into meta- or macro-modules. Instead, the system provides a number of highly advanced modules, some of which implement extremely complex synthesis techniques. For example, the system includes a module which alone is almost equivalent to the Yamaha DX7 synthesiser: it contains a six-operator FM synthesis architecture with thirty-two preset algorithms (see Chapter 3).

Figure 2.9 The *Edit window* for the *Oscillator* module. Each module is provided with default parameter values which can be edited via its *Edit window*

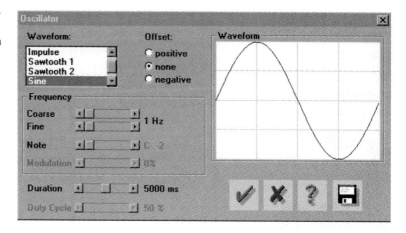

Since there is no efficient provision for playing an instrument via a controller or a score file, the parameters for each module must previously be set beforehand. Each module is provided with default parameter values which will be implemented if the user does not choose to edit them. Such parameters are edited via the *Edit window*, which is opened by double-clicking on the module's icon in the workspace. Figure 2.9 illustrates the *Edit window* for the *Oscillator* module. This module generates a periodic signal using different waveforms chosen from a menu; the signal can be a sinewave, a square wave, a sawtooth wave, a triangular wave or a pulse wave.

The complete list of modules is located to the left of the workspace and a module is selected simply by clicking on its name. When the mouse is moved over the workspace area an icon for the selected module automatically appears, and it can be placed anywhere within the area. The modules of the system are divided into three categories: *Generators, Processes* and *Analyses*.

Generators are modules which create different types of signals, which are either *audio signals* (e.g. oscillator and noise generator) or *control signals* (e.g. envelope generator). Control signals are used to control the synthesis parameters of other modules; for example the cut-off frequency of a resonant filter. There is a variety of generators: some of them are basic synthesis units (e.g. oscillator and envelope) but most of them encompass an entire synthesis algorithm (e.g. FM, FOF and Karplus–Strong Algorithm; see Chapters 3, 4 and 6).

The *Processes* category includes a number of modules to modify specific attributes of a given sound. Processes are by their very nature required to have at least one input and an output, with

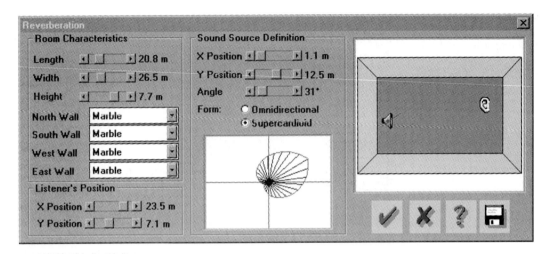

Figure 2.10 The *Reverberation* module has the ability to model the characteristics of various room acoustics, such as the dimensions of the room and absorption properties of the wall's materials

the exception of the *Output* module, which is always the last component of an instrument. This category adds all the capabilities of a powerful sound editor and sampling system to the synthesis facilities of the system. Indeed, it embraces a number of signal processors and effect units, including a powerful reverb module with the ability to model the characteristics of various room acoustics, such as the dimensions of the room and absorption properties of the wall's materials (Figure 2.10).

The modules of the *Analysis* category analyse the signal passed through them without changing it in any way. Their purpose is solely to generate a visual representation of the analysis data. The available analysis techniques include STFT (see Chapter 6) and sonogram.

2.2.4 GrainWave

GrainWave is a synthesis programming system for Power Macintosh/Power PC computers designed by the composer Michael Berry, at the Mills College Center for Contemporary Music in Oakland, California, USA. It works on the standard Power PC setup with no need for additional hardware or sound cards.

In essence, the modus operandi of GrainWave is similar to that of Virtual Waves or Audio Architect. In GrainWave, instruments are built by combining unit generators called *operators* (Figure 2.11). There are four classes of operators, namely: signal generators, controls, mixers and effects (Figure 2.12).

In order to design an instrument, the operators must be listed on four different tab panels, one for each class of operators

Figure 2.11 In GrainWave, instruments are built by combining operators

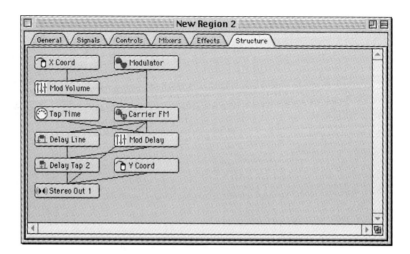

Figure 2.12 The *Feedback FM* signal generator of GrainWave

(Figure 2.13). The operators are then networked via pop-up menus that display all possible connections for a particular variable. Every variable has its own pop-up menu, which is opened by clicking the mouse on it.

In GrainWave parlance, an instrument is referred to as a *region* and various regions can be put together to form a *patch* on a patch window. The patch is the basic document of GrainWave. It contains one or more regions, all the lookup tables (called *FTables* in GrainWave jargon) needed by the regions and, eventually, sound files. Regions can be played in real time using the mouse, the computer keyboard or via MIDI.

Lookup tables can be edited graphically or mathematically using the *FTable Editor*, a facility that provides various editing tools such as selection, cut and paste operation, and a pencil to draw lines and curves (Figure 2.14).

GrainWave uses Opcode's Open MIDI System (OMS) for MIDI communication. OMS is freely available on the Internet; refer to the system's documentation for more details. Like other synthesis programming systems available on the CD-ROM for Windows 95, GrainWave can also run alongside another program, such as a sequencer, on the same Macintosh machine. Indeed, Michael Berry devised Max Virtual Controller, a stand-

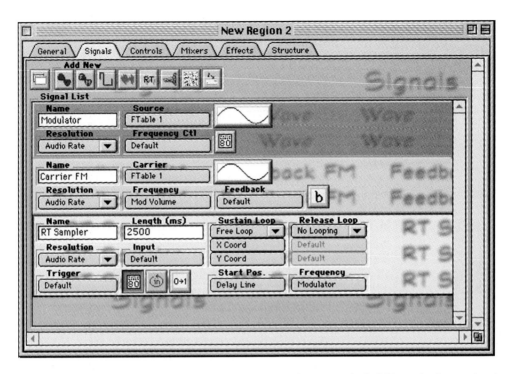

Figure 2.13 The *Signals* tab panel. There are four classes of operators in GrainWave: signal generators (or simply signals), controls, mixers and effects. In order to design an instrument, the operators must be listed on four different tab pab panels, one for each class of operator

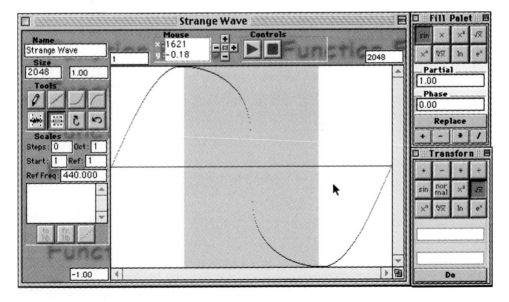

Figure 2.14 The *FTable Editor* provides various editing tools to create and modify waveforms

alone application, to control GrainWave patches. The Max Virtual Controller application, a collection of examples and a comprehensive tutorial are also available on the accompanying CD-ROM.

2.3 Software synthesiser

2.3.1 Reality

Reality is a software synthesiser that runs on PC-compatible computers developed by Seer Systems, Los Altos, USA. The modus operandi of Reality is not rooted in the concept of assembling instruments by interconnecting synthesis building blocks. Instead, Seer Systems ventured to implement a large multipurpose instrument that integrates different synthesis techniques. As a rough comparison, imagine Reality as the combination of an analog synthesiser, a sampler and a Yamaha VL-type of machine, integrated onto one single MIDI instrument accompanied by a comprehensive patch editor. The major drawback of this approach is that Reality limits the electronic musician to the synthesis techniques that it can produce, but this is the price one

Figure 2.15 The main screen of Reality provides access to three main working areas of the system: *Bankset*, *Program* and *Options*

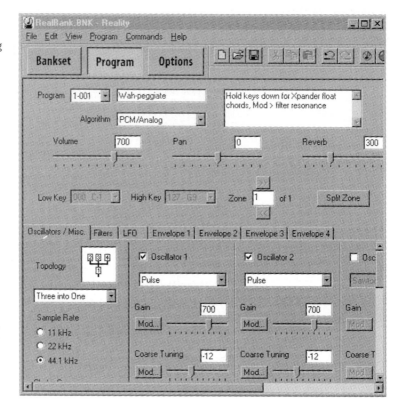

has to pay for high-level systems of any kind. Nevertheless, Reality is easy to utilise and it is much quicker to learn its usage than learning actual programming languages, such as CLM, or programming systems, such as GENERATOR. Reality is flexible for building its 'patches' and it can hold more than one thousand patches in its memory. Also, it offers 64-voice polyphony and is multi-timbral to sixteen MIDI-channels.

The system is structured as follows:

• At the highest level is the *bankset*: a storage structure that can hold up to 1336 programs, arranged in groups of 128 (because of the MIDI way of enumerating units, i.e. from 0 to 127).

• A *program*, which may be a *patch* or a *patchwork*, is more or less equivalent to a timbre on a standard MIDI synthesiser. Programs can be selected from any MIDI controller able to send program change messages; see Rumsey (1994) for an introduction to MIDI. A patch is simply the set-up for a timbre using any of the synthesis capabilities of the system, whereas a patchwork may combine different types of timbres.

Figure 2.16 The *Options* view is used to set up a number of global playback tools, including the master volume, a reverb, a chorus and MIDI source

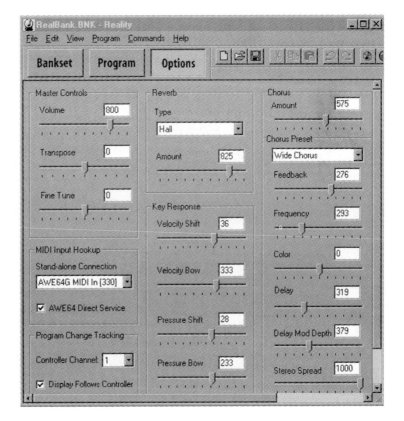

Reality provides two main groups of synthesis facilities, referred to as *PCM/analog settings* and *algorithmic methods*. The former includes a number of settings for implementing synthesis techniques inspired by analog synthesisers, such as FM and additive synthesis (see Chapters 3 and 6). In this case, there are four oscillators that can be configured in a number of topologies, four envelope generators, four filters and a number of controllers that indeed resemble the functioning of an analog synthesiser. The oscillators can produce a number of different waveforms, including sawtooth, triangle, pulse and sinusoid; oscillators can also take in sound samples to act as a template waveform. As for the *algorithms methods*, the program provides settings for a number of synthesis methods inspired by physical modelling techniques (see Chapter 4).

The main screen of the program provides access to the three main working areas of the system: the *Bankset* view, where one selects programs and sampled wave templates; the *Program* view (Figure 2.15), where the editable controls of a sound are located; and the *Options* view, where one can set up a number of global playback utensils such as the master volume, a reverb, a chorus and the MIDI source (Figure 2.16).

2.4 Non-generic synthesis software

2.4.1 Som-A

Som-A is a system for additive synthesis (see Chapter 6) devised by Aluizio Arcela at the University of Brasília, Brazil (Arcela, 1994). The word 'som' in Portuguese means sound and 'soma' means addition.

At the heart of *Som-A* is an elementary but smart programming language for the specification of additive instruments and scores; both are saved to a single file denominated by Aluizio Arcela as a *spectral chart*. The system provides a front-end for operation which includes an editor for writing spectral charts and facilities for playback.

The language itself is extremely simple; it has only one unit generator, called *H-unit* (Figure 2.17). An instrument is nothing more than a set of H-units, where each unit is set up to produce a partial of the desired spectrum.

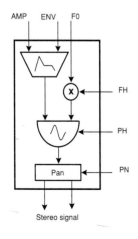

Figure 2.17 The H–unit is the only unit generator of Som-A

Each *H-unit* contains a sinusoidal oscillator, an envelope and a panning control, and each of them requires the specification of the following parameters:

- The shape of envelope and duration (ENV)
- Amplitude (AMP)
- Reference frequency; i.e. fundamental frequency (F0)
- Partial produced; in order to define the actual frequency produced (FH)
- Phase of the sinusoid in degrees (PH)
- Panning (PN)

The language has only five different instructions and its syntax is somewhat similar to Lisp; commands are lists whose first element is the instruction. Som-A instructions are *VAL, INS, EXE, STP* and *FIM* and all should appear in a spectral chart (see Som-A's user manual for more details):

- VAL sets initialisation values such as timing in seconds and sampling rate
- INS used to define an instrument: life-span in seconds (i.e. for how long it will run), its name and a list of H-units; each unit is defined as follows (harmonic, phase, (envelope), panning)
- EXE indicates the beginning of the score section
- STP indicates the end of the score section
- FIM indicates the end of the spectral chart

The overall structure of a spectral chart is as follows:

```
(VAL <header parameters>)
(INS <instrument definition>)
(EXE <timing information>)
       (<sound event 1>)
       (<sound event 2>)
          ...
       (<sound event n>)
(STP)
(FIM)
```

As an example, consider the following spectral chart:

```
(VAL 0 6 44100)
(INS 6 instr1
       (1 0 ((0 0) (1000 255) (0 511)) 1)
       (5 45 ((0 0) (1000 255) (0 511)) 0) )
(EXE 0 6)
       (instr1 0 1 370.0 20)
       (instr1 1 1 425.3 20)
       (instr1 2 2 466.2 20)
       (instr1 4 2 622.3 20)
(STP)
(FIM)
```

41

The first instruction notes that the chart will produce a sound file of 6 seconds duration at a sampling rate of 44.1 kHz. The second instruction defines one instrument, called *instr1*, whose life-span is equal to 6 seconds. The instrument produces two partials, the fundamental and the fifth harmonic. The fundamental is panned to the right channel and the fifth harmonic panned to the left. The fifth harmonic is 45 degrees out of phase in relation to the fundamental and the envelope ((0 0) (1000 255) (0 511)) is used for both partials. The lines between instructions EXE and STP plays a sequence of four pitches: 370.0 Hz, 425.3 Hz, 466.2 Hz and 622.3 Hz, respectively. The first two pitches last for 1 second each whereas the remaining two pitches each last for 2 seconds.

Although the version of Som-A on the CD-ROM runs only on PC-compatible computers under Windows 95, by the time you read this book the platform-independent Java version of Som-A may already be available from the University of Brasília via the Internet.

2.4.2 Diphone

Diphone, designed by Xavier Rodet and his team at Ircam in Paris, France, is a program for making transitions between sounds; an effect commonly referred to as *sound morphing*. The program runs on Macintosh computers and is part of a set of programs supplied to the subscribers of the Ircam Forum. Originally, the term 'diphone' refers to a transition between two phonemes, but for musical purposes this notion has been expanded to mean the transition between any two sounds, not necessarily of vocal origin.

Ircam's Diphone addresses one of the most challenging synthesis problems of both artificial speech and sound composition: the concatenation of sound sequences. In computer-synthesised speech the algorithm used to utter a specific syllable, such as 'mu', may not produce satisfactory results on different words. For instance, take the words 'music' and 'mutation'. If a computer produces the syllable 'mu', as for the word 'music', at the beginning of the word 'mutation' we promptly feel that the latter does not sound right. This is because the transition (i.e. the *diphone*) between 'u' and 's' in the word 'music' and the transition between 'u' and 't' in the word 'mutation' involve different spectral behaviour. This transition problem is also evident in music composition where certain articulations and musical passages are clearly more appropriate to our ear than others. This problem is even more challenging for electronic musicians because they deal with a much larger repertoire of sounds to combine and articulate.

Although this book does not focus on music composition, Diphone is introduced here as an example of a synthesis system that was primarily designed to address a problem pertinent to all electronic musicians, and more specifically to those working with sampled sounds and sound montage; e.g. *acousmatic music* (Bayle, 1993; Norman, 1996). Composers working with recorded sound normally concatenate the sounds using splicing or cross-fading techniques (see Chapter 3). In most cases, however, the results do not sound satisfactory because the inner contents of the sounds involved do not always match. Diphone attempts to render this task more effective by using analysis and resynthesis techniques (see Chapter 6).

Each sound sample in Diphone is submitted to an analysis stage that extracts crucial information on how its spectrum evolves in time. This analysis provides a 'multi-parametric' representation of the sound; it contains information about its fundamental frequency (i.e. pitch), plus the frequency, the amplitude and the phase of each component of its spectrum. The advantage of such representation over straight sampling is that this information can be manipulated individually and mapped to the parameters of a synthesis algorithm suited for resynthesising the sound; for example, *additive synthesis* (discussed in Chapter 6). If the analysis information is not changed, then the outcome of the resynthesis process should be equal to the original sound. Conversely, if the analysis information is changed, then the resulting resynthesis will sound different. For example, if the fundamental frequency is changed, then the result will be a sound at a different pitch. Note, however, that this modification is by no means equivalent to playing back the sound at a faster or slower speed, as samplers do. In this case, the timbre may be considerably changed, whereas in Diphone the timbre of the sound would remain intact.

Each sound in Diphone is referred to as a *segment*. The program produces a set of analysis data for each segment and stores them on a *dictionary* of segments. Diphone provides an intuitive user

Figure 2.18 Diphone concatenates the sounds by applying an algorithm that interpolates the analysis data of neighbouring segments

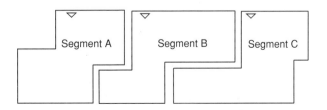

interface for operation, in which one can drag segments from a dictionary and drop them onto a working area, referred to as the *sequence window* (Figure 2.18). At the top of the sequence window the computer displays the segments for concatenation and at the bottom it provides a menu of 'parameters' for monitoring the concatenation.

The program concatenates the sounds by applying an algorithm that interpolates the analysis data of neighbouring segments. It is important to note that Diphone does not manipulate the sounds directly, only the analysis data. The whole sequence is synthesised only when the user commands the computer to do so.

Each segment is represented by an icon and it has three distinct areas: a central area and two adjacent interpolating areas (Figure 2.19). The user can adjust the extent of the interpolating areas and define the type of interpolation algorithm for each neighbourhood; e.g. linear, logarithmic, etc.

One of the major strengths of Diphone is the possibility to edit and monitor the concatenation graphically. The bottom area of the sequence window lists the parameters that will be interpolated and other global settings, such as a transposition index and a gain factor for amplification or attenuation of the signal. Here one can pull down graphic break-point functions to monitor the

Figure 2.19 Each segment is represented by an icon and it has three distinct areas: a central area and two adjacent interpolating areas

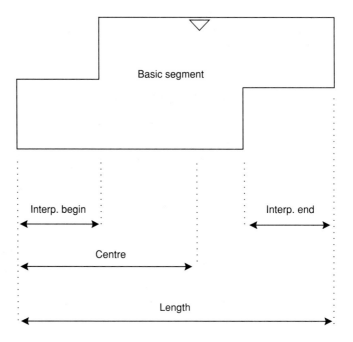

Figure 2.20 Concatenation parameters can be monitored graphically before and after the actual interpolation takes place

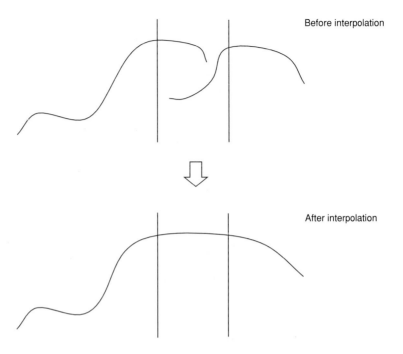

Before interpolation

After interpolation

behaviour of these parameters, with options to visualise the plot before and after the interpolation takes place (Figure 2.20). Furthermore, there are facilities for editing the plot manually with a 'pencil'.

Although originally designed to work with additive synthesis, Diphone also supports other synthesis techniques such as formant synthesis (discussed in Chapter 6) and physical modelling (Chapter 4).

2.4.3 The CDP sound transformation toolkit

The CDP sound transformation toolkit is part of a large package called *The CDP Computer Music Workstation* currently available for the Atari Falcon, PC-compatible and Silicon Graphics platforms. CDP stands for Composer's Desktop Project and it was created in the mid-1980s in York by a pool of British composers and sound engineers with the objective of producing affordable computer systems for music composition. At a time when computer music equipment could be sustained only by moneyed institutions, the CDP team managed to manufacture a fairly powerful system that worked on Atari ST computers provided with DAC/ADC hardware manufactured in-house and a PCM converter. The CDP system soon became very popular among composers and educators in Britain and elsewhere.

The CDP package is formed by a number of individual programs, each dedicated to a specific task. The package is subdivided into three groups of programs: (1) Synthesis, (2) Groucho Signal Processing and (3) Spectral Sound Transformation. The first group includes Csound, a Music N-style synthesis programming language that resembles pcmusic in many ways, but the last two groups constitute CDP's most powerful and unique pieces of software.

The Groucho Sound Processing group includes programs for manipulating the samples of a sound file in the time-domain. There are more than 100 different programs in this group, which allow for many ways to manipulate a sound, including all the time-domain modelling techniques described in Chapter 5.

As for the Spectral Sound Transformation group, it includes dozens of highly specialised programs for spectral manipulation, which constitute one of the most powerful toolkits of its kind. They deal directly with the timbre or tone colour of a sound, thus making possible the modification of timbre in unprecedented ways. The Spectral Sound Transformation tools carry out a spectral analysis of the sound to user-defined degrees of complexity, whereupon the numerical data can be manipulated in many different ways (see Chapter 6).

2.4.4 LASy

LASy (Linear Automata Synthesis) is a program made specifically for cellular automata lookup table synthesis. It was designed by Jacques Chareyron (1990), at the University of Milan's 'Laboratorio di Informatica Musicale' (LIM), Italy. LASy runs on Macintosh computers and it is part of a pool of programs developed at LIM, which together form a larger integrated computer music workbench called *Intelligent Music Workstation*, or IMW.

The cellular automata lookup table synthesis technique is discussed in Chapter 4. LASy basically works by applying a cellular automaton to a lookup table containing an initial waveform. At each playback cycle of the lookup table, the cellular automaton algorithm processes the waveform. The intention is to let the samples of the lookup table be in perpetual mutation, but according to a sort of 'genetic code'.

Considering that LASy's synthesis technique does not demand heavy computation to process the samples, sounds can be produced in real time via a MIDI controller, mouse or even the computer's alphanumeric keyboard. The program provides facilities for configuring the MIDI communication.

Figure 2.21 LASy instruments are defined in terms of an initial waveform, a rule for the cellular automaton and envelopes for amplitude and pitch

TabOnde 0	
Wave Tab 0	
1	*Wave 1
2	Wave 2
3	Wave 3
4	Wave 4
5	Wave 5
6	Wave 6
7	Wave 7
8	Wave 8
9	Wave 9
10	Wave 10
11	Wave 11
12	Wave 12

TabIns 0	
Ins.Tab 0	
1	Ins 1
2	Ins 2
3	Ins 3
4	Ins 4
5	Ins 5
6	Ins 6
7	Ins 7
8	Ins 8
9	Ins 9
10	Ins 10
11	Ins 11
12	Ins 12

TabRule 0	
Rule Tab 0	
1	*Rule 1
2	Rule 2
3	Rule 3
4	Rule 4
5	Rule 5
6	Rule 6
7	Rule 7
8	Rule 8
9	Rule 9
10	Rule 10

TabEnv 0	
Env.Tab 0	
1	*Env 1
2	Env 2
3	Env 3
4	Env 4
5	*Env 5
6	Env 6
7	Env 7
8	Env 8
9	Env 9
10	Env 10

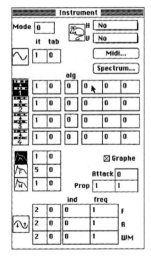

Figure 2.22 The main window for the specification of LASy synthesis parameters

LASy instruments are defined in terms of an initial waveform, a rule for the cellular automaton (referred to as the *transition rule*) and envelopes for amplitude and pitch. The system provides tools for the specification of these components individually so that the user can build his or her own library of waveforms, rules and envelopes (Figure 2.21). Instruments are either created by combining components selected from these libraries or from scratch on a window where the user can set up all these parameters at once (Figure 2.22).

Initial waveforms are created within the system itself either by using a break-point function or by adding up sinusoids, but sampled sounds may also be used. A great variety of instruments can be created by combining different waveforms and transition rules.

LASy has the ability to synthesise a large variety of sounds with diverse spectral evolutions, particularly sounds with fast transients at the very beginning of the sound. The program is particularly good for producing wind-like and plucked strings-like sounds. Yet, the ingredient that still makes LASy unique is its ability to synthesise unusual sounds but with some resemblance to the real acoustic world.

Jacques Chareyron classifies the output of LASy into three main groups, according to the type of cellular automata rules employed (Chareyron, 1990):

- *Sounds with simple evolution leading to a steady-state ending*: simple cellular automata rules (refer to the system's documentation) generate monotonous evolution of the sound spectrum, where the spectral envelope follows either an increasing or decreasing curve, leading to a steady-state ending.
- *Sounds with simple evolution but with no ending*: by increasing the complexity of the cellular automata rules one obtains endless successions of similar but not completely identical waveforms.
- *Everlasting complex sounds*: the more complex the rules, the more unpredictable is the behaviour of the cellular automata, and therefore the more complex the evolution of the sound

This classification is, of course, very general inasmuch as the boundaries between the categories are crudely vague. Nevertheless, as LASy's author himself suggests, they are a good starting point for further experimentation.

2.4.5 Wigout and TrikTraks

Wigout and TrikTraks form a system for PC-compatible platforms (DOS and Windows) specifically designed for sequential waveform composition (see Chapter 5). The system was designed by the composer Arun Chandra in collaboration with researchers of the School of Music at the University of Illinois, USA (Chandra, 1994).

Like Som-A, the heart of both Wigout and TrikTraks are two elementary programming languages for the specification of waveforms and transitions. Each of them has its own sound compiler.

Wigout works by generating sequences of sample segments that, taken together, compose a waveform. It can generate complex combinations of waveforms of arbitrary length, limited only by the amount of memory on the computer (see Chapter 5 for more details about the sequential waveform composition technique).

In Wigout, a sound is specified in terms of three separate files referred to as *segments*, *states* and *events*. The *segments* file contains descriptions for the segments that will be used to compose the waveform. Each segment is described in terms of amplitude and duration (measured in samples). In addition, a segment also has a range (maximum and minimum values) and an increment of change. It can have three different shapes: square, triangular or curved, referred to as a *wiggle*, a *twiggle*, or a *ciggle*, respectively. A segment is defined by three lines of code. For example (comments are always preceded by the symbol '#'):

```
a1 wiggle              # identifier, type
10 20 1 1              # initial, max, min, inc (duration)
0 32000 -32000 4000    # initial, max, min, inc (amplitude)
```

The first line declares a wiggle segment labelled as *a1*. Then, the second line determines the duration of the wiggle. The first item on this line is its initial duration, followed by its maximum duration, minimum duration, and increment. All values are given in samples. In this case, the initial duration is ten samples, the maximum duration it will reach is twenty samples, and the minimum duration it will reach is one sample. On each iteration of this segment, the duration will change by one sample. Finally, the third line describes the amplitude of *a1*: its initial value, followed by its maximum value, minimum value, and increment. In this case, the initial amplitude will be 0 but it will increase to 32 000 and decrease to –32 000, and will change by 4000 on each iteration. Every time this segment is iterated, its duration and its amplitude will increase by their respective increments. Once they reach their maximum ranges, they will start shrinking and when they reach their minima, they will start increasing again.

The *states* file contains a specification of sequences of segments that make up a particular combination, or state, for generating waveforms. For example:

```
# id segments
s1 a1 a2 a3
s2 a2 a3 a2 a3 a4
```

In this case there are two states, labelled as *s1* and *s2*, respectively. The first state consists of three segments: *a1* followed by *a2*, followed by *a3*, and the second state consists of five segments: *a2* followed by *a3*, followed by *a2* again, followed by *a3* again, followed by *a4*. A state is iterated from its start time until its end time, according to the specifications given in the events file. On every iteration, each of its segments changes its duration and its amplitude.

The *events* file contains a list of states, their start and end times (in seconds), and their stereo location (a number between 0 and 1):

s1 0 5 0 # state label , start time, end time, stereo location
s2 0 5 1

In this example, state *s1* (as defined in the states file) begins at time 0 seconds, iterates until time 5 seconds, and will be played on the left of the stereo field (left = 0). Similarly, state *s2* also begins at 0 seconds and iterates until 5 seconds, but it will be played of the right of the stereo field (right = 1).

TrikTraks complements Wigout by providing the means for the specification of *transformational paths* during the production of a waveform. Here the electronic musician can specify a number of different functions to change the waveform, including sine, triangular and polynomial trajectories.

2.4.6 SMS

SMS is an analysis and resynthesis program (see Chapter 6) designed by Xavier Serra, of the Pompeu Fabra University in Barcelona, Spain. The program on the accompanying CD-ROM

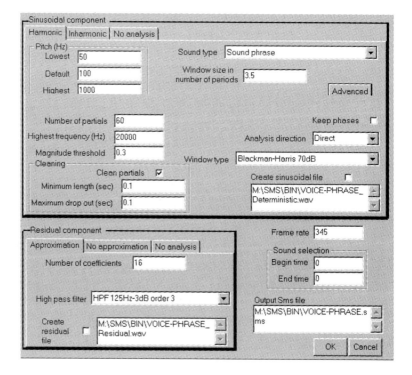

Figure 2:23 SMS is a powerful analysis and resynthesis program for spectral modelling synthesis

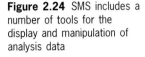

Figure 2.24 SMS includes a number of tools for the display and manipulation of analysis data

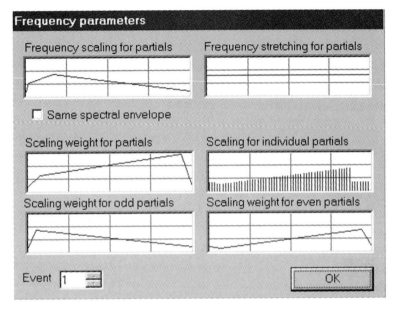

runs on PC-compatible computers under Windows 95 and it has been developed as a graphical front-end (Figure 2.23) for a powerful analysis and resynthesis consort, originally available as a set of individual programs that were activated manually via DOS or Unix command lines.

The new Windows-based front-end integrates the whole SMS consort into one program and includes a number of additional tools for the display and manipulation of the analysis data, as well as graphical facilities for the specification of resynthesis parameters (Figures 2.24 and 2.25). The program offers gadgets for three main types of tasks: *analysis, transformation-resynthesis* and *event-list control*. Whilst the analysis group involves tools for displaying a sound and the analysis data generated by the system, the transformation-resynthesis bunch includes facilities for manipulating the analysis data for resynthesis. Finally, the event-list control tools allow for detailed control of resynthesis whereby the musician can define tasks such as spectral interpolation of different sounds.

Although SMS is an abbreviation for Spectral Modelling Synthesis, it is important to bear in mind that in this book the author uses the term 'spectral modelling' to refer to a class of synthesis techniques and not to a specific program. The SMS program on the CD-ROM in fact embodies the *resynthesis by reintegration of discarded components* technique described in Chapter 6. For an in-depth discussion about the inner function-

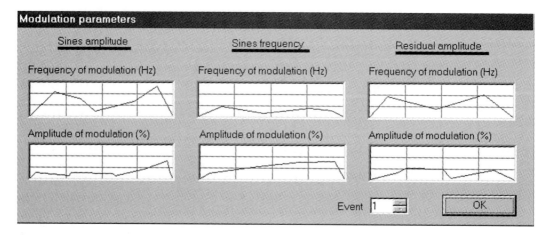

Figure 2.25 The *Modulation parameter* window allows for the specification of modulation parameters for resynthesis

ing of SMS the reader should consult the documentation provided with the program.

2.4.7 Chaosynth

Chaosynth is a granular synthesis program that uses cellular automata to control the production of the sounds; granular synthesis is discussed in Chapter 5. This program is a version for PC-compatible computers, implemented by Joe Wright from an earlier version designed by the author in the early 1990s.

The granular synthesis of sounds involves the production of thousands of short sonic particles that are combined to form larger complex sound events (Figure 2.26). This synthesis technique is inspired by the British physicist Dennis Gabor's famous proposition that large, complex sound events are composed of streams of simple acoustic particles (Gabor, 1947). Norbert Wiener, one of the pioneers of Cybernetics, also adopted a similar concept to measure the information content of a sonic message (Wiener, 1964). It was the composer Iannis Xenakis (1971), however, who suggested one of the first applications of granular sound representation for sound composition purposes. Since then, a few granular synthesis systems have been proposed but, so far, most of these systems use stochastic methods to control the production of the sonic particles. Chaosynth proposes a different method: the use of cellular automata (CA).

CA are mathematical models of dynamic systems in which space and time are discrete and quantities take on a finite set of discrete values. CA are often represented as a regular array with a variable at each site, metaphorically referred to as a cell. The state of the CA is defined by the values of the variables at each

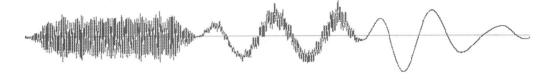

Figure 2.26 Granular synthesis of sounds involves the production of thousands of short sonic particles that are combined to form larger, complex sound events

cell. The whole automata evolves according to an algorithm, called a *transition function*, that determines the value of each cell based on the value of their neighbourhood. As implemented on a computer, the cells are represented as a grid of tiny rectangles whose values are indicated by different colours (Figure 2.27).

A wide variety of CA and transition functions have been invented and adapted for many modelling purposes in many scientific areas, including physics, computing, biology and meteorology. CA have also attracted the interest of musicians because of their organisational principles. Various composers and researchers have used CA to aid the control of both higher-level musical structures or musical form (Miranda, 1993), and lower-level structures, such as the spectra of individual sound events. Chaosynth uses CA to control the lower-level structure of sounds.

Figure 2.27 Most granular synthesis systems use stochastic methods to control the production of the sonic particles, but Chaosynth uses cellular automata. The state of the cellular automata is defined by the values of the variables at each cell. As implemented on a computer, the cells are represented as a grid of tiny rectangles whose values are indicated by different colours

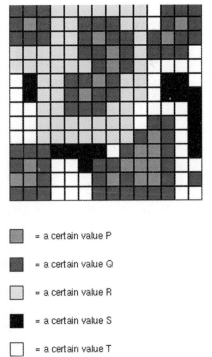

■ = a certain value P

■ = a certain value Q

▢ = a certain value R

■ = a certain value S

☐ = a certain value T

The program plus more information about its inner functioning, as well as comprehensive system documentation and examples, can be found in folder *Chsynth* on the accompanying CD-ROM.

2.4.8 Morph

Morph is an analysis and resynthesis program devised by Raffaele de Tintis, at the University of Milan's 'Laboratorio di Informatica Musicale' (LIM), Italy. It runs on PC-compatible computers under Windows 95. Morph is a work-in-progress program that embodies Raffaele de Tintis' ongoing research work on sound hybridisation techniques; an essay about his research can be found on the CD-ROM within Morph's materials.

The program's role is basically to fuse two sounds in order to create a new compound sound. For example, you can use Morph to fuse a bell sound with a flute sound in order to create a sound with perceptual characteristics of both in it. However simple this task may appear to be, Morph does not work by simply mixing both audio signals. In order to create convincing hybrid sounds one must actually 'mix' the spectra of the source sounds.

Figure 2.28 The main cross-synthesis panel of Morph, where the electronic musician can specify connections between the spectral components of the source sounds

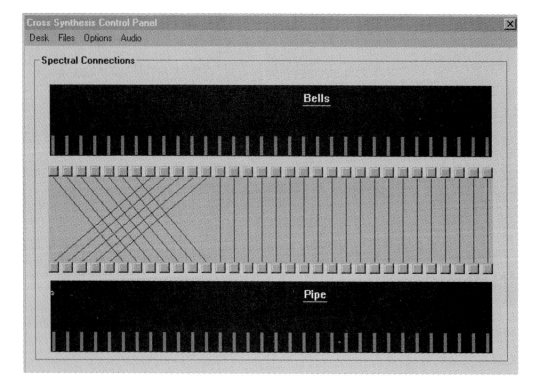

Raffaele de Tintis distinguishes between two main types of spectral fusion: *cross-synthesis* and *hybridisation*. Cross-synthesis is a method whereby one employs the spectral envelope of one sound to shape the spectrum content of another. The main limitation of cross-synthesis is that it does not allow for the control of the spectral fusion. It works solely upon raw numerical operations, disregarding any high-level perceptual features of the two sound sources. The mere cross-synthesis of spectral components is often inefficient for spectral fusion, because different sounds may have properties that are more suitable than others for transformation. Conversely, hybridisation employs spectral fusion methods that take into account higher-level perceptual properties of the source spectra; it supports the formulation of strategies for spectral fusion based upon perceptual features of both sounds (e.g. type of attack, resonant properties, etc.).

Morph offers facilities for experimenting with the two methods of sound fusion: cross-synthesis and hybridisation. On the cross-synthesis front, Morph offers an unusual facility whereby musicians can change the order of the 'connections' between the spectral components of the source sounds. The main cross-synthesis control panel (Figure 2.28) portrays the spectra of the

Figure 2.29 The main hybridisation panel of Morph allows for the specification of a number of parameters for spectral fusion

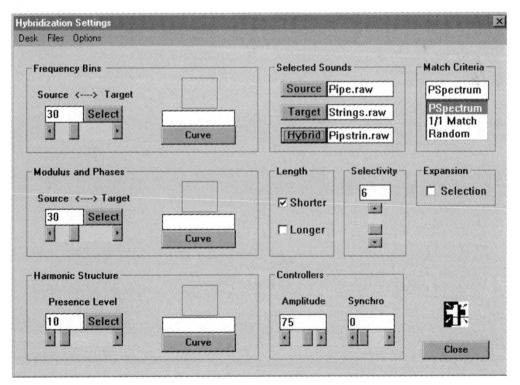

source sounds divided into thirty-two bands, as for a graphic equaliser. Here the musician can specify connections between the bands of the top and bottom rows so that the envelope of the bands at the top will be reshape the spectrum of the bands at the bottom row. Changing these connections will not produce better cross-synthesis results but will certainly create interesting effects. For example, linking the high-frequency bands of one sound with the low-frequency bands of another will produce a new sound with rising high-frequency components, before the appearance of low frequencies; an effect that goes against the acoustic laws of most acoustic musical instruments. As for hybridisation, Morph offers a number of controllers for customising the spectral fusion (Figure 2.29).

3 Loose modelling techniques

Loose modelling techniques inherited much of their ethos from analog synthesis. At the beginning of electronic music in the 1950s, composers ventured to synthesise sounds by superimposing a few sinusoidal waves. Although very exciting to begin with, at the end of the day electronic musicians realised that in order to fulfil their increasing quest for more sophisticated sounds they would need to stockpile many more sinusoids than they had expected. Moreover, it had become evident that in order to achieve sounds with a minimum degree of acoustic coherence, each sinusoid would need to follow its own course in terms of amplitude, frequency and phase. The technical hindrance from producing good additive synthesis at the time forced composers to look for alternative synthesis methods.

In this context, techniques such as amplitude modulation (AM), frequency modulation (FM) and waveshaping arose and immediately gained ground in the electronic music studio. Composers were delighted to discover that they could create rich time-varying spectra with far fewer resources than if they were to produce everything using the additive method alone. However, there was a price to pay: whereas it was straightforward to understand additive synthesis parameters such as amplitude and frequency, unfamiliar parameters such as FM's modulation index and index of deviation were not so easy to control. Indeed, as will become evident in this chapter, the only way to describe the behaviour of the resulting spectrum in these cases is by means of abstract mathematical formulations; hence the origin of the term 'loose model'.

On the whole, loose modelling techniques tend to be very easy to implement, but their synthesis parameters are not straightforward to control because they bear little relation to the acoustic world. Computer technology also inspired a few loose modelling techniques, such as Walsh synthesis and synthesis by binary instruction.

3.1 Amplitude modulation

Modulation occurs when some aspect of an audio signal (called a *carrier*) varies according to the behaviour of another audio signal (called a *modulator*). Amplitude modulation therefore occurs when a modulator drives the amplitude of a carrier. One of the pioneers of amplitude modulation in music was the composer Karlheinz Stockhausen in the 1960s (Maconie, 1976).

The *tremolo* effect may be considered to be a starting point example of amplitude modulation; it is achieved by applying a very slow sub-audio rate of amplitude variation to a sound (i.e. less than approximately 18 Hz). If the frequency of the variation is raised to the audible band (i.e. higher than approximately 18 Hz) then additional partials (or *sidebands*) will be added to the spectrum of the signal.

Simple amplitude modulation synthesis uses only two sinewave generators (or oscillators): one for the carrier and the other for the modulator. The frequency of the carrier oscillator is usually represented as f_c whilst the frequency of the modulator oscillator is represented as f_m.

Figure 3.1 In classic AM, the output of the modulator oscillator is added to an offset amplitude-value

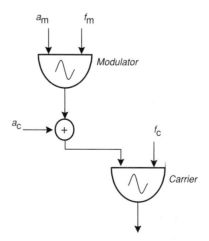

Complex amplitude modulation may involve more than two signals; for example, the amplitude of *oscillator C* is modulated by the outcome of *oscillator B*, which in turn is amplitude modulated by *oscillator A*. Signals other than sinewaves (e.g. noise) may also be employed for either carriers or modulators. The more complex the signalling system, the more difficult it is to predict the outcome of the instrument. This book focuses only on simple amplitude modulation.

There are two variants of amplitude modulation: *classic amplitude modulation* (AM) and *ring modulation* (RM).

3.1.1 Classic amplitude modulation

In classic amplitude modulation (AM) the output from the modulator is added to an offset amplitude value (Figure 3.1). Note that if there is no modulation, the amplitude of the carrier would be equal to this offset value. The amplitude of the modulator is specified by an amount of the offset amplitude value in relation to a modulation index.

If the modulation index is equal to zero then there is no modulation, but if it is higher than zero then the carrier wave will take an envelope with a sinusoidal variation (Figure 3.2). In simple AM, the spectrum of the resulting signal contains energy at three frequencies: the frequency of the carrier plus two sidebands, one below and the other above the carrier's frequency value. The values of the sidebands are established by subtracting the frequency of the modulator from the carrier and by adding the frequency of the modulator to the carrier, respectively. The amplitude of the carrier frequency remains unchanged, whilst

Figure 3.2 The amplitude of the carrier signal is controlled by the modulator signal

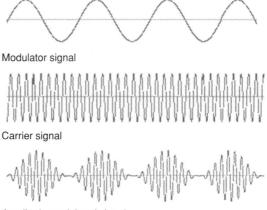

Modulator signal

Carrier signal

Amplitude modulated signal

Figure 3.3 In simple AM, the spectrum of the resulting signal contains energy at the frequency of the carrier plus two sidebands

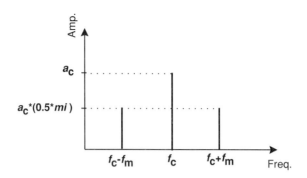

the amplitudes of the sidebands are calculated by multiplying the amplitude of the carrier by half of the value of the modulation index, represented as *mi* (see Appendix 1 for detailed mathematical specifications). For example, when *mi* = 1, the sidebands will have 50 per cent of the amplitude of the carrier (Figure 3.3).

3.1.2 Ring modulation

In ring modulation (RM) the amplitude of the carrier is determined entirely by the modulator signal. Thus if there is no modulation, then there is no sound (Figure 3.4). In simple RM (i.e. when both signals are sinewaves), the resulting spectrum contains energy only at the sidebands ($f_c - f_m$ and $f_c + f_m$); the frequency of the carrier wave will not be present. RM therefore may distort the pitch of the carrier signal. For instance, if f_c = 440 Hz and f_m = 110 Hz, then the instrument will produce two sidebands of 330 Hz and 550 Hz, respectively. In RM the energy of the modulator signal is split between the two resulting sidebands (Figure 3.5). Because there is no fundamental frequency in the resulting spectrum, the sounds of RM usually do not have a strong sensation of pitch.

Ring modulation may also be achieved by the multiplication of two signals (Figure 3.6). The multiplication of two sounds results in a spectrum containing frequencies that are the sum of and difference between the frequencies of each component in the first sound and those of each component in the second.

Both AM and RM can use signals other than sinusoids, applying the same principles discussed above. In any case, great care must be taken in order to avoid aliasing distortion (i.e. generation of frequencies above 50 per cent of the sampling rate) because the highest frequencies of the two sounds will be additive.

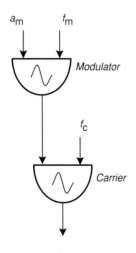

Figure 3.4 In RM the amplitude of the carrier signal is entirely determined by the modulator

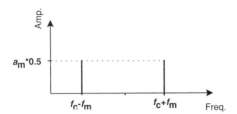

Figure 3.5 In RM, the frequency of the carrier will not be present and the amplitude of the modulator is split between the two resulting sidebands

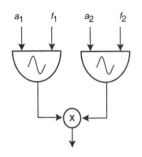

Figure 3.6 The multiplication of two signals is also a form of RM

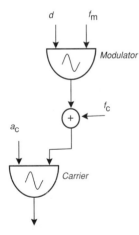

Figure 3.7 The most basic FM configuration is composed of two oscillators, called modulator and carrier oscillators. The output of the modulator is offset by a constant value and the result is applied to control the frequency of the carrier

3.1.3 Amplitude modulation examples

An example of a simple AM instrument (*Am.mae*) implemented using Audio Architect can be found on the CD-ROM (in folder *Aamets* in the Audio Architecture materials). An example is also available in GrainWave for Macintosh (in folder *Patches*, in the GrainWave materials).

A collection of examples in pcmusic are available in folders *Tutor* and *Ringmod*. The reader will also find a number of annotated examples on various aspects of AM and RM in Robert Thompson's pcmusic tutorial, from page 55 to page 65.

3.2 Frequency modulation

Synthesis by frequency modulation (FM) was originated in the late 1960s by the Stanford University composer John Chowning (Chowning, 1973; Chowning and Bristow, 1986). FM synthesis is based upon the same principles used for FM radio transmission. Chowning's first experiments employed an audio signal (called a *modulator*) to control the frequency of an oscillator (called a *carrier*) in order to produce spectrally rich sounds.

The *vibrato* effect may be considered as a starting point example to illustrate the functioning of FM. The fundamental difference, however, is that vibrato uses a sub-audio signal to modulate the carrier. A sub-audio signal is a low-frequency signal, well below the human hearing threshold which is approximately 20 Hz. The resulting sound, in this case, has a perceptibly slow variation in its pitch. If the modulator's frequency is set to a value above the human hearing threshold, then a number of partials are added to the spectrum of the carrier's sound.

There are a number of variations in FM instrument design. The most basic of the FM instruments is composed of two oscillators, called *modulator* and *carrier* oscillators, respectively (Figure 3.7). This simple architecture is capable of producing a surprisingly rich range of distinctive timbres. More complex FM instruments may employ various modulators and carriers, combined in a number of ways.

3.2.1 Simple frequency modulation

Figure 3.7 illustrates a simple frequency modulation architecture. The output of the modulator is offset by a constant, represented as f_c, and the result is then applied to control the frequency of the carrier. If the 'amplitude' of the modulator is equal to zero, then there is no modulation. In this case, the output from the carrier will be a simple sinewave at frequency f_c. Conversely, if the 'amplitude' of the modulator is greater than zero, then modulation occurs and the output from the carrier will be a signal whose frequency deviates proportionally to the 'amplitude' of the modulator. The word amplitude is within quotation marks here because this parameter has, in fact, a special name in FM theory: *frequency deviation* (represented as d), and its value is usually expressed in Hz.

Figure 3.8 If the frequency deviation (i.e. the amplitude of the modulator signal) is gradually increased, then the period of the carrier's output increasingly expands and contracts, proportional to the value of the frequency deviation

The parameters for the FM instrument portrayed in Figure 3.7 are as follows:

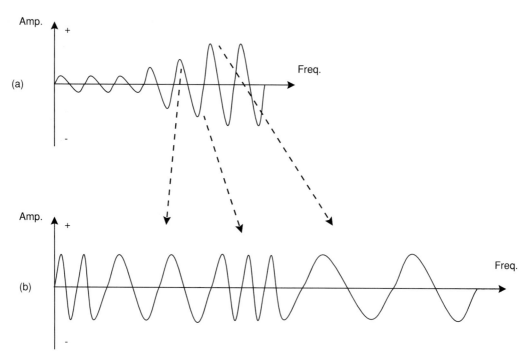

d = frequency deviation
f_m = modulator frequency
a_c = carrier amplitude
f_c = offset carrier frequency

The role of frequency deviation and of the modulator frequency is illustrated in Figure 3.8. Both parameters can drastically change the form of the carrier wave. If the modulator frequency is kept constant whilst increasing the frequency deviation, then the period of the carrier's output will increasingly expand and contract, proportional to the frequency deviation. If the frequency deviation remains constant and the modulator frequency is increased, then the rate of the deviation will become faster.

3.2.2 The spectrum of FM sounds

The simplicity of its architecture and its capability to produce a great variety of different timbres made FM synthesis more attractive than other techniques available at the time of its invention. Moreover, it does not necessarily need waveforms other than sinusoids, for both modulator and carrier, in order to produce interesting musical sounds.

Calculating the frequencies of the partials

The spectrum of an FM sound is composed of the offset carrier frequency (f_c) and a number of partials on either side of it, spaced at a distance equal to the modulator frequency (f_m). The partials generated on each side of the carrier frequency are usually called sidebands. The sideband pairs are calculated as follows: $f_c + k \times f_m$ and $f_c - k \times f_m$ where k is an integer, greater than zero, which corresponds to the order of the partial counting from f_c (Figure 3.9).

The amplitudes of partials are determined mostly by the frequency deviation. When there is no modulation (i.e. $d = 0$) the

Figure 3.9 The spectrum of an FM sound is composed of the offset carrier frequency and a number of partials on either side of it, spaced at a distance equal to the modulator frequency

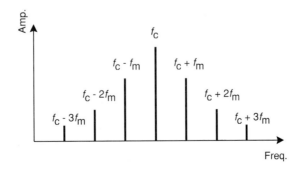

63

Figure 3.10 As the modulation index increases, the number of audible partials also increases and the energy of the offset carrier frequency is distributed among them

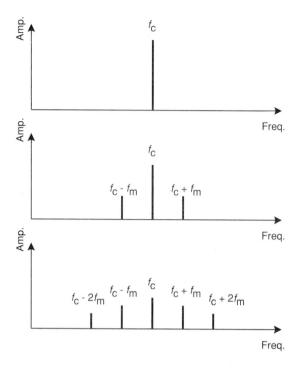

power of the signal resides entirely in the offset carrier frequency f_c. Increasing the value of d produces sidebands at the expense of the power in f_c. The greater the value of d, the greater the number of generated partials and, therefore, the wider the distribution of the power between the sidebands.

The FM theory provides a useful tool for the control of the number of audible sideband components and their respective amplitudes: the *modulation index*, represented as i. The modulation index is the ratio between the frequency deviation and the modulator frequency: $i = d/f_m$.

As the modulation index increases from zero, the number of audible partials also increases and the energy of the offset carrier frequency is distributed among them (Figure 3.10). The number of sideband pairs with significant amplitude can generally be predicted as $i + 1$; for example, if $i = 3$ than there will be four pairs of sidebands surrounding f_c.

Calculating the amplitudes of the partials

The offset carrier frequency may often be the most prominent partial in an FM sound; in this case it will define the pitch of the sound. Sound engineers tend to refer to the carrier frequency as the fundamental frequency, but musicians often avoid this nomenclature because in music the term 'fundamental frequency' is normally associated with pitch and in FM the carrier frequency does not always determine the pitch of the sound.

The amplitudes of the components of the spectrum are determined by a set of functions known as *Bessel functions* and represented as $B_n(i)$. An in-depth mathematical study of Bessel functions is beyond the scope of this book; we will only introduce the basics in order to understand how they determine the amplitude of the partials of an FM-generated sound.

Figure 3.11 shows the graphical representation of four Bessel functions: $B_0(i)$, $B_1(i)$, $B_2(i)$ and $B_3(i)$, respectively. They determine amplitude scaling factors for pairs of sidebands, according to their position relative to the offset carrier's frequency. Note that these are not absolute amplitude values, but scaling factors.

The carrier amplitude (a_c) usually defines the overall loudness of the sound, but the amplitudes for individual partials are calculated by scaling the given carrier amplitude according to the factors established by the Bessel functions. For instance, $B_0(i)$ determines the scaling for the offset carrier frequency (f_c), $B_1(i)$ the scaling for the first pair of sidebands ($k = 1$), $B_2(i)$ the scaling for the second pair of sidebands ($k = 2$), and so on.

Figure 3.11 Bessel functions determine the amplitude scaling factors for pairs of sidebands, according to their position relative to the offset carrier's frequency

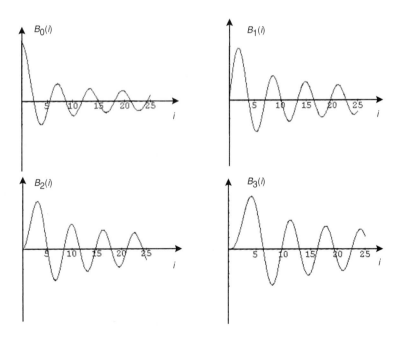

Figure 3.12 The vertical axis of a Bessel function plot indicates the amplitude scaling factor according to the value of the modulation index represented by the horizontal axis. For example, the amplitude scaling factor for the second pair of sidebands will be value 0.11 if $i = 1$; i.e. $B_2(1) = 0.11$

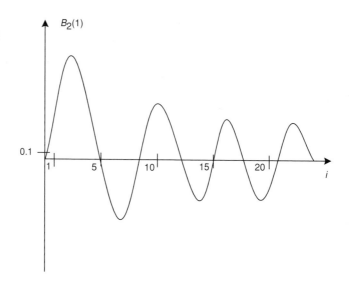

The vertical axis indicates the amplitude scaling factor according to the value of the modulation index represented by the horizontal axis. For example, when $i = 0$ (i.e. no modulation) the offset carrier frequency will sound at its maximum factor (i.e. 1) and the amplitudes of all sidebands will be zero; if $i = 1$, then the scaling factor for f_c will decrease to approximately 0.76, the factor for the first pair of sidebands will have a value of 0.44, the second pair will have a value of 0.11, etc. (Figure 3.12). The value of the modulation index must be large in order to obtain significant amplitudes in high-order sidebands. See Appendix 1 for a list of scaling factors.

One important rule to bear in mind when calculating an FM spectrum is that the scaling factors for the odd partials on the left of the sideband pair are multiplied by –1.

Figure 3.13 Negative amplitudes indicate sidebands that are out of phase and may be represented by plotting it downwards in the frequency axis

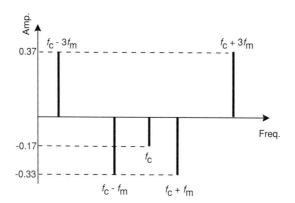

Dealing with negative amplitudes

In addition to the above rule, notice that Bessel functions indicate that sidebands may have either positive or 'negative' amplitude, depending on the modulation index. For example, if $i = 5$, the scaling factor for the first pair of sidebands will be approximately −0.33. In fact, 'negative' amplitude does not exist. The negative signal here indicates that the sidebands are out of phase. This may be represented graphically by plotting it downwards on the frequency axis (Figure 3.13), but the phase of a partial does not produce an audible effect unless another partial of the same frequency happens to be present. In this case,

Figure 3.14 When the offset carrier frequency is low and the modulation index is set very high, the modulation will almost certainly produce sidebands that fall in the negative frequency domain of the spectrum (a). As a rule, negative sidebands will fold around the 0 Hz axis and mix with the sidebands in the positive domain. Note, however, that reflected partials will reverse their phase and add algebraically to the partials of the same values that were present before the reflection (b). Phase information does not necessarily need to be represented in the resulting plot (c)

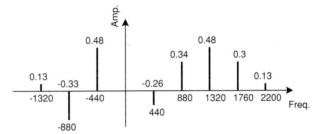

(a)

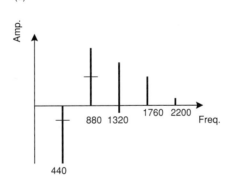

(b)

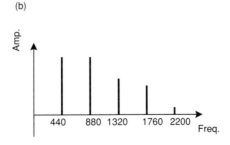

(c)

the amplitudes of the conflicting partials will either add or subtract, depending on their respective phases.

Dealing with negative frequencies and the Nyquist distortion

An important phenomenon to consider for the calculation of an FM spectrum occurs when the offset carrier frequency is very low and the modulation index is set high. In this case, modulation may produce sidebands that fall in the negative frequency domain of the spectrum. As a rule, all negative sidebands will fold around the 0 Hz axis and mix with the sidebands in the positive domain. Reflected partials from the negative domain will, however, reverse their phase. For instance, the settings $f_c =$ 440 Hz, $f_m =$ 440 Hz and $i = 3$, produce the following negative sidebands: −1320 Hz, −880 Hz, and −440 Hz. These partials fold into the positive domain, with reversed phase, and add algebraically to the partials of the same values that were there before the reflection; Figure 3.14 illustrates this phenomenon. In fact, phase information does not necessarily need to be represented in the resulting spectrum, because people do not hear the effect of phase inversion.

Another phenomenon that is worth remembering here is that partials falling beyond the Nyquist limit (see Chapter 1) fold over and reflect into the lower portion of the spectrum.

3.2.3 Synthesising time-varying spectra

The ability to provide control for time-varying spectral components of a sound is of critical importance for sound synthesis. The amplitudes of the partials produced by most acoustic instruments vary through their duration. They often evolve in complicated ways, particularly during the attack of the sound. This temporal evolution of the spectrum cannot be heard explicitly at all times. Occasionally, the evolution might occur over a very short time span or the whole duration of the sound itself may be very short. Even so, it establishes an important cue for the recognition of timbre.

Frequency modulation offers an effective parameter for spectral evolution: the *modulation index* (*i*). As has been already demonstrated, the modulation index defines the number of partials in the spectrum. An envelope can thus be employed to time-vary the modulation index in order to produce interesting spectral envelopes that are unique to FM. Note, however, that by linearly increasing the modulation index the instrument does not necessarily increase the power of high-order sidebands linearly.

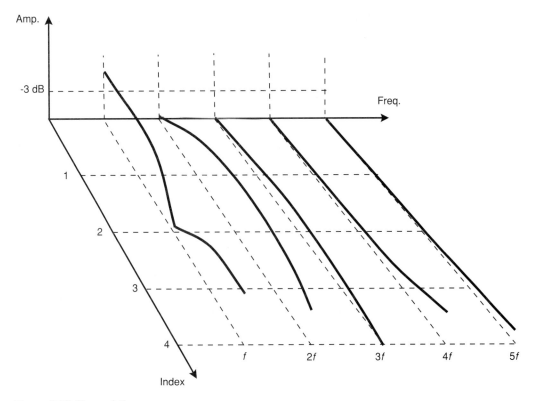

Figure 3.15 The evolution of each partial is determined by its corresponding Bessel function. A specific partial may increase or decrease its amplitude according to the slope of its Bessel function. Note, however, that linearly increasing the modulation index does not necessarily increase the amplitude of the high-order sidebands linearly

Remember that the evolution of each partial is determined by its corresponding Bessel function. A specific partial may therefore increase or decrease its amplitude, according to the slope of its Bessel function at specific modulation values (Figure 3.15).

3.2.4 Frequency ratios and sound design

Timbre control in FM is governed by two simple ratios between FM parameters. One is the ratio between the frequency deviation and the modulator frequency and has already been introduced: it defines the modulation index (i). The other is the ratio between the offset carrier frequency and the modulator frequency, called *frequency ratio* and represented as $f_c{:}f_m$. The frequency ratio is a useful tool for the implementation of a phenomenon that is very common among conventional instruments, that is, achieving variations in pitch whilst maintaining the timbre virtually unchanged.

If the frequency ratio and the modulation index of a simple FM instrument are maintained constant but the offset carrier frequency is modified then the sounds will vary in pitch, but their timbre will remain unchanged. In this case, it is much easier to think in terms of frequency ratios rather than in terms of values for f_c and f_m separately. For example, whilst it is clear to see that $f_c = 220$ Hz and $f_m = 440$ Hz are in ratio 1:2, when presented with the figures $f_c = 465.96$ Hz and $f_m = 931.92$ Hz, it is not so obvious.

As a rule of thumb, frequency ratios should always be reduced to their simplest form. For example, 4:2, 3:1.5 and 15:7.5 are all equivalent to 2:1. Basic directives for sound design in terms of these simpler ratios are given as follows:

Case 1: if f_c is equal to any integer and f_m is equal to 1, 2, 3 or 4, then the resulting timbre will have a distinctive pitch, because the offset carrier frequency will always be prominent.

Case 2: if f_c is equal to any integer and f_m is equal to any integer higher than 4, then the modulation produces harmonic partials but the fundamental may not be prominent.

Case 3: if f_c is equal to any integer and f_m is equal to 1, then the modulation produces a spectrum composed of harmonic partials; e.g. the ratio 1:1 produces a sawtooth-like wave.

Case 4: if f_c is equal to any integer and f_m is equal to any even number, then the modulation produces a spectrum with some combination of odd harmonic partials; e.g. the ratio 2:1 produces a square-like wave.

Case 5: if f_c is equal to any integer and f_m is equal to 3, then every third harmonic partial of the spectrum will be missing; e.g. the ratio 3:1 produces narrow pulse-like waves.

Case 6: if f_c is equal to any integer and f_m is not equal to an integer, then the modulation produces non-harmonic partials; e.g. 2:1.29 produces a 'metallic' bell sound.

3.2.5 Composite frequency modulation

Composite FM involves two or more carrier oscillators and/or two or more modulator oscillators. There are a number of possible combinations and each of them will create different types of spectral compositions. On the whole, complex FM produces more sidebands but the complexity of the calculations to predict the spectrum also increases.

There are at least five basic combinatory schemes for building composite FM instruments:

1 Additive carriers with independent modulators
2 Additive carriers with one modulator
3 Single carrier with parallel modulators
4 Single carrier with serial modulators
5 Self-modulating carrier

Figure 3.16 Composite FM using two carriers with independent modulators

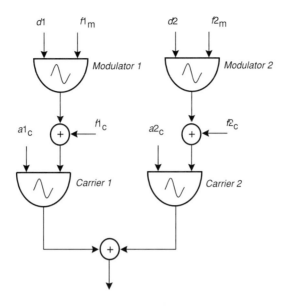

Additive carriers with independent modulators

This scheme is composed of two or more simple FM instruments working in parallel (Figure 3.16). The spectrum therefore is the result of the addition of the outputs from each instrument.

Additive carriers with one modulator

This scheme employs one modulator oscillator to modulate two or more carrier oscillators (Figure 3.17). The resulting spectrum is the result of the addition of the outputs from each carrier oscillator.

Single carrier with parallel modulators

This scheme employs a more complex signal to modulate a carrier oscillator: the result of two or more sinewaves added together (Figure 3.18). In this case, the formula for the calculation of a simple FM spectrum is expanded in order to accommodate multiple modulator frequencies and modulation indices. For example, in the case of two parallel modulator oscillators, the sideband pairs are calculated as follows:

$$f_c - (k_1 \times f_{m1}) + (k_2 \times f_{m2})$$
$$f_c - (k_1 \times f_{m1}) - (k_2 \times f_{m2})$$
$$f_c + (k_1 \times f_{m1}) + (k_2 \times f_{m2})$$
$$f_c + (k_1 \times f_{m1}) - (k_2 \times f_{m2})$$

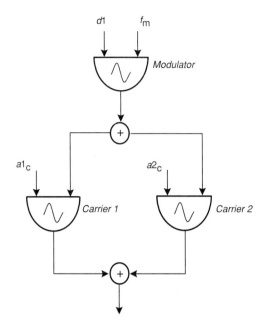

Figure 3.17 Composite FM using two carriers and a single modulator

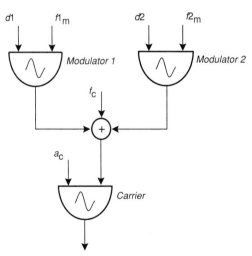

Figure 3.18 Composite FM using a single carrier with two parallel modulators

This formula looks complicated but, in fact, it simply states that each of the partials produced by one modulator oscillator (i.e. $k_1 \times f_{m1}$) forges a 'local carrier' for the other modulator oscillator (i.e. $k_2 \times f_{m2}$). The larger the number of parallel modulators, the greater the amount of nested 'local carriers'. The amplitude scaling factors here result from the multiplication of the respective Bessel functions: $B_n(i_1) \times B_m(i_2)$. A practical example is given in Appendix 1.

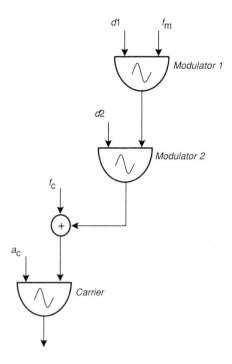

Figure 3.19 Composite FM using a single carrier with two modulators in series

Single carrier with serial modulators

This scheme also employs a complex signal to modulate a carrier oscillator. In this case, however, the modulating signal is a frequency modulated signal (Figure 3.19). The sideband frequencies are calculated using the same method used above for parallel modulators, but the calculation of the amplitude scaling factors is different. The 'order' of the outermost modulator is used to scale the modulation index of the next modulator: $B_n(i_1) \times B_m(n \times i_2)$.

The main differences between the spectrum generated by serial modulators and parallel modulators, using the same frequency ratios and index, are that:

1   The former tends to have sidebands with higher amplitude values than the latter
2   No sideband components from $B_m(i)$ are generated around the carrier centre frequency; e.g. $B_0(i_1) \times B_1(0 \times i_2) = 0$.

Self-modulating carrier

The self-modulating carrier scheme employs the output of a single oscillator to modulate its own frequency (Figure 3.20). The

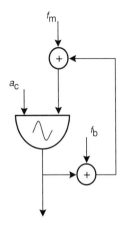

Figure 3.20 The self–modulating carrier scheme employs the output of a single oscillator to modulate its own frequency

oscillator output signal is multiplied by a *feedback factor* (represented as f_b) and added to a frequency value (f_m) before it is fed back into its own frequency input; f_b may be considered here as a sort of modulation index.

This scheme will always produce a sawtooth-like waveform due to the fact that it works with a 1:1 frequency ratio by default; that is, the modulation frequency is equal to its own frequency. The amplitudes of the partials increase proportionally to f_b. Beware, however; as this parameter is very sensitive, values higher than $f_b = 2$ may lead to harsh white noise.

The self-modulating carrier scheme is sometimes preferable to a simple 1:1 FM instrument. The problem with simple FM is that the amplitudes of its partials vary according to the Bessel functions, but this variation is not linear. The number of sidebands increases by augmenting the modulation index, but their amplitudes do not rise linearly. This grants an 'unnatural' colouration to the sound which may not always be desirable. The amplitudes of the partials produced by a self-modulating oscillator increase more linearly according to the feedback factor (f_b).

3.2.6 Modular implementation

There are various approaches for implementing FM synthesisers; for example, a computer programming-based approach can be substantially different from a commercial MIDI-keyboard approach. Whilst the former may have highly flexible tools for instrument design, the latter may not provide this flexibility but can be much easier to operate. In general, the industry of MIDI-keyboard type of synthesisers tends to produce better 'interfaces' than the academic programming languages, but at the expense of flexibility. Yamaha is a well-known trademark for commercially successful FM synthesisers and its success is probably due to an ingenious industrial procedure: basic FM mechanisms are encapsulated into higher-level modules called *operators*. In this case, the user only has access to a limited set of parameters to control these operators; for example, the modulation index is not explicitly available for manipulation.

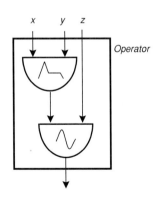

Figure 3.21 The operator of a DX7 type synthesiser basically consists of an envelope generator and an oscillator and it can be used either as a carrier or a modulator

An operator basically consists of an envelope generator and an oscillator, and it can be used as either a carrier operator or a modulator operator (Figure 3.21). Synthesisers are programmed by appropriately interconnecting a number of operators to form what is referred to as an *algorithm*. Various instruments are then defined by specifying appropriate parameter values for specific topologies of algorithms (Figure 3.22). An algorithm plus a set

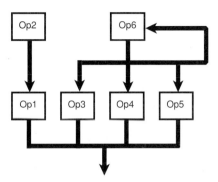

Figure 3.22 Some FM synthesisers are programmed by appropriately interconnecting a number of operators to form different algorithms

of definite parameter values constitutes a *patch*. A collection of patches is then provided in read-only memory (ROM) but the user can edit the parameter values and the algorithm in order to create his or her own patches. Depending on the model of the synthesiser, four or six operators are provided and can be combined into a number of different algorithms.

3.2.7 FM examples

On the accompanying CD-ROM, Kenny McAlpine's Audio Architect tutorial (in folder *Audiarch*) illustrates how to implement a simple FM instrument. Also, Nicky Hind's CLM tutorial (in folder *Clm*) presents a step-by-step introduction to implementing FM instruments in CLM. A number of FM examples are also available in pcmusic, in folder *Fm* in the pcmusic materials.

An interesting approach to FM instrument design is given by Virtual Waves (in folder *Virwaves*). Virtual Waves holds a module called *F.M. Synthesis* which embodies the modular

Figure 3.23 The *F.M. Synthesis* module of Virtual Waves embodies the modular implementation of FM

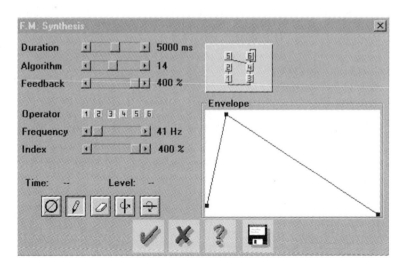

Figure 3.24 In this example, the sound starts noisily and then the partials suddenly settle within a certain formant configuration

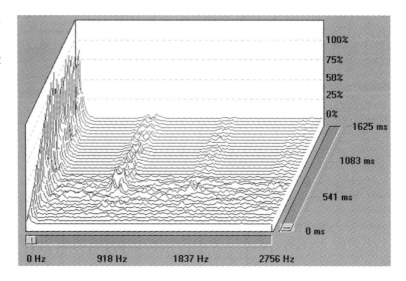

implementation of FM. The module has six operators which can be combined into thirty-two different algorithms. The *Edit* window for this module provides the means for selecting an algorithm and for setting up the parameters for each operator (Figure 3.23). A few examples can be found in folder *Synthvir*; for instance, *Fm3.syn* produces a sound that starts with random configuration, then the partials settle quickly within certain formant regions (Figure 3.24).

For those wishing to experiment with feedback FM, GrainWave provides *Feedback FM*, an operator especially designed for single modulator feedback frequency modulation; an example is provided in folder *Patches*, in the GrainWave materials on the CD-ROM.

3.3 Waveshaping synthesis

Waveshaping synthesis (also termed *non-linear distortion* or *non-linear processing*) creates composite spectra by applying distortion to a simple sound (e.g. a sinusoid). The technique functions by passing a sound through a unit that distorts its waveform, according to a user-specified directive. In waveshaping synthesis jargon, the distortion unit is called *waveshaper* and the user-specified directive is called *transfer function*. The first experiments with this synthesis technique were carried out by the composer Jean-Claude Risset in the late 1960s.

As a metaphor to understand the fundamentals of waveshaping synthesis, imagine that Figure 3.25(a) represents a note played on

Figure 3.25 If the volume of a note played on an electric stringed instrument connected to a vacuum-tube amplifier (a) is increased to maximum, the vacuum-tubes will be saturated and the sound will clip (b). If the amplitude of the note is increased at its origin, before entering the amplifier (c), then the output will clip even more (d)

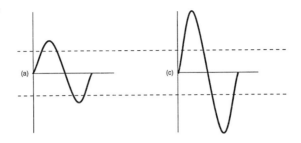
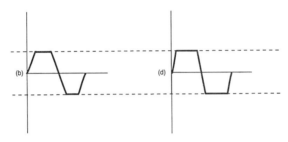

an electric stringed instrument connected to a vacuum-tube amplifier. If the volume knob of the amplifier is increased to its maximum, the vacuum-tubes will be saturated and the sound will clip (Figure 3.25(b)) and if the amplitude of the note is increased at its origin, before entering the amplifier (Figure 3.25(c)), then the output will clip even more (Figure 3.25(d)). If the note is a sinusoid, as portrayed in Figure 3.25, the louder the input, the more squared the output wave will be. If the note has a complex spectrum, then the output will be a signal blurred by distortion.

Figure 3.26 In waveshaping, the richness of the resulting spectrum (b) is proportional to the amplitude of the input signal (a)

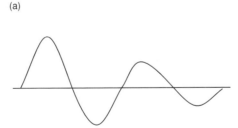

Next, assume that the instrument produces a sustained note, gradually losing amplitude, as portrayed in Figure 3.26. In this case, the output will be a sound whose amplitude remains almost constant, but the waveform will vary continuously.

Amplitude sensitivity is one of the key features of waveshaping synthesis. In the above example, the amount of amplification (or distortion) is proportional to the level of the input signal. The spectrum of the output therefore becomes richer as the level of the input is increased.

3.3.1 The waveshaper

The crucial element of the waveshaping technique is the *waveshaper* (Figure 3.27): a processor that can alter the shape of the waveform passing through it.

In a linear processor, such as an *ideal amplifier,* a change in the amplitude of the input signal produces a similar change in the output signal, but the shape of the waveform remains unchanged. For example, doubling the amplitude of the input signal will double the amplitude of the output. Conversely in a non-linear processor, the relationship between input and output signals depends upon the amplitude of the input signal and the nature of the non-linearity. In most cases, non-linear processors modify the waveform of the input signal. This modification generally results in an increase in the number and intensity of its partials.

The waveshaper is characterised by *a transfer function* which relates the amplitude of the signal at the output to the signal at

Figure 3.27 The waveshaper is a processor that can alter the shape (i.e. form) of the waveform passing through it

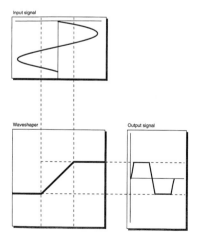

Figure 3.28 A graphic representation of a transfer function where the amplitude of the input signal is plotted on the horizontal axis and the output on the vertical

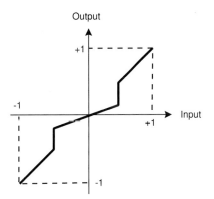

the input. For example, Figure 3.28 shows a graphic representation of a transfer function where the amplitude of the input signal is plotted on the horizontal axis and the output on the vertical. For a given input value, the output of the waveshaper can be determined by computing the corresponding output value on the graph of the transfer function. If the transfer function is a straight diagonal line, from −1 to +1, the output is an exact replica of the input. Any deviation from a straight diagonal line introduces some sort of modification.

The success of a waveshaping instrument depends upon the choice of a suitable transfer function. Drawing transfer functions intuitively would be one way to experiment with waveshaping, but it might become difficult to predict the results systematically. Fortunately, there are a number of tools for constructing waveshapers out of mathematical functions that allow for some degree of prediction of the resulting spectra; for instance, *polynomials* and *trigonometric functions*. To a great extent the best transfer functions are described using polynomials. See Appendix 1 for detailed mathematical specifications.

3.3.2 Chebyshev polynomials and spectra design

A particular family of polynomials called *Chebyshev polynomials of the first kind* has been widely used for specifying transfer functions for waveshaping synthesis. Chebyshev polynomials are represented as follows: $T_k(x)$ where k represents the order of the polynomial and x represents a sinusoid. Chebyshev polynomials have the useful property that when a cosine wave with amplitude equal to one is applied to $T_k(x)$, the resulting signal is a sinewave at the kth harmonic. For example, if a sinusoid of amplitude equal to one is applied to a transfer function given by the seventh-order Chebyshev polynomial, the result will be a sinusoid at seven times the frequency of the input. Chebyshev polynomials for T_1 to T_{10} are given in Appendix 1.

Because each separate polynomial produces a particular harmonic of the input signal, a certain spectrum composed of various harmonics can be obtained by summing a weighted combination of Chebyshev polynomials, one for each desired harmonic (see Appendix 1 for an example).

Given the relative amplitudes of the required harmonics, most sound synthesis programming languages and systems provide facilities to plot Chebyshev-based transfer functions on a waveshaper. On most systems, the waveshaper is implemented as a table, or array, whose size is compatible with the maximum amplitude value of the input signal. For instance, to fully waveshape a sinusoid oscillating between –4096 and +4096 sampling units, the transfer function should span a table containing 8193 values (i.e. 4096 for the negative samples, plus 4096 for the positive, plus 1 for the zero crossing). Note that the expression 'amplitude 1' has been conventionally used on a number of occasions here to refer to the maximum input amplitude value that a waveshaper can manage.

3.3.3 Distortion index and time-varying spectra

Variations on the amplitude of the input signal can activate different 'portions' of the waveshaper. Waveshaping synthesis is therefore very convenient to synthesise sounds with a considerable amount of time-varying spectral components.

The amplitude of the input signal is often associated with a *distortion index*, controlled by an envelope. In this case, changes in volume will cause changes in the spectrum. This coincides in theory with the behaviour of a number of acoustic instruments, where louder playing produces more overtones.

3.3.4 Waveshaping examples

Figure 3.29 shows a simple waveshaping instrument implemented using GENERATOR; try *Waveshap1.ens* in folder *Ensembs* in the GENERATOR materials on the accompanying CD-ROM. The waveshaper here is a 'clipper' programmed to clip when the amplitude of the input sinewave is outwith the interval –0.5 and +0.5. The amplitude of the sinewave is controlled by a fader and it varies from 0 to 1. Values up to 0.5 produce a clean sinewave, but as the amplitude ascends above 0.5, the sound is increasingly clipped thus producing cumulatively richer spectra. Some examples in pcmusic and GrainWave are also available on the CD-ROM.

Figure 3.29 An example of a simple waveshaping instrument in GENERATOR

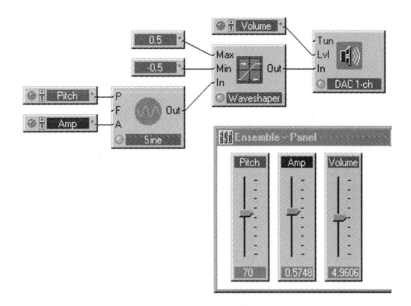

3.4 Walsh synthesis

Whereas AM, FM and waveshaping are synthesis techniques inherited from the world of analog synthesis, Walsh synthesis is an inherently digital technique. Walsh synthesis works based upon square waves instead of sinewaves; it creates other waveforms by combining square waves. Inasmuch digital systems work with only two numbers at their most fundamental level (i.e. 0 and 1), square waves are very straightforward to manipulate digitally.

It is curious to note that although Walsh synthesis may resemble the additive synthesis method of sinewave summations, this technique is based upon a very different paradigm. Walsh synthesis produces waveforms from the summation of Walsh functions rather than from the summation of partials. Walsh functions are measured in *zero crossings per second* (zps) instead of *cycles per second* (cps or Hz) and they are not explicitly related to specific 'harmonics' (Figure 3.30).

Although is it possible to predict the spectrum of Walsh synthesis by means of complex mathematical calculations, this technique is not efficient for methods rooted in Fourier's theory. Actually, the most difficult waveform to produce with Walsh synthesis is the sinewave, which is the fundamental unit of Fourier's theory. Conversely, the most difficult waveform to produce by summing sinewaves is the square wave, which is the fundamental unit of Walsh synthesis.

Figure 3.30 Walsh functions are the backbone of Walsh synthesis

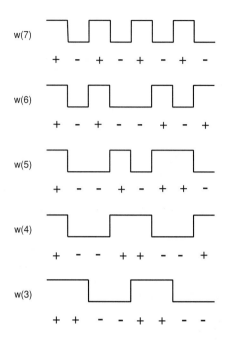

Despite its affinity with the digital domain, Walsh synthesis has not been sufficiently studied and composers have rarely used it. There is great potential, however, for an entirely new theory for sound synthesis based on Walsh functions; for example, sounds could be analysed using a Walsh-based technique and resynthesised using fast Walsh transform (FWT). Also, modulation involving Walsh functions may be able to produce certain types of waveforms much more efficiently than modulation using sinusoids. A few technical articles about Walsh synthesis can be found in various issues of *Journal of the Audio Engineering Society*; e.g. Hutchins (1973, 1975).

An example of a simple instrument inspired by the Walsh synthesis technique can be found on the accompanying CD-ROM in folder *Ensembs* in the GENERATOR materials. The instrument (*Walsh.ens*) uses three multistep oscillators to produce the waveforms corresponding to the first three Walsh functions portrayed in Figure 3.30. GENERATOR is particularly suited for producing Walsh-type waveforms because of its unique multistep oscillator modules.

3.5 Binary instruction

Synthesis by binary instruction is a curious technique developed in the 1970s at the Institute of Sonology in the Netherlands. This technique is commonly known as *non-standard synthesis*, but in this book we prefer to use the more specific term – binary synthesis – in order to distinguish it from other methods that could also be classified as non-standard; for example, Walsh synthesis and sequential waveform composition (discussed in Chapter 5).

Most synthesis techniques presented in this book have their roots, in one way or another, in metaphorical models seeking inspiration from the acoustic world. The great majority of systems for software synthesis actually bias the design of instruments towards this type of inspiration. The binary instruction technique diverges from this approach. It departs from the idiomatic use of low-level computer instructions, with no attempt to model anything acoustic.

Standard systems for software synthesis normally generate samples from a relatively high-level specification of acoustically related components and parameters, such as oscillators, envelope, frequency and amplitude. In binary instruction, samples are generated from the specification of low-level computer instructions with no reference to any pre-defined synthesis paradigm. The sound is described entirely in terms of digital processes. The rationale for this radical approach is that the sounds should be produced using the low-level 'idiom' of the computer. If there is such a thing as a genuine 'computer timbre', then binary instruction would be the most qualified technique to synthesise it.

Binary instruction functions by using basic computer instructions such as logical functions and binary arithmetic to process sequences or binary numbers. The result of the processing is output as samples through an appropriate digital-to-analog converter (DAC). The compelling aspect of this technique is the speed in which samples are processed. Since there is no need to compile or interpret high-level instructions, sounds can be produced in real time on very basic machines. This was a very great achievement for the 1970s.

Different sounds are associated with different programs coded in the assembler language of the computer at hand. The idea is interesting but it turned out to be a damp squib because assembler is a very low-level language and musicians would hardly have a vested interest in this level of programming. In order to alleviate the burden of writing assembler programs to produce sounds, Paul Berg and his colleagues at the Institute of Sonology developed a language called PILE (Berg, 1979) for the specification of binary instruction instruments. Similarly, at the University of Edinburgh, Stephen Holtzmann developed a sophisticated system for generating binary instruction instruments inspired by research in artificial intelligence (Holtzman, 1978).

3.6 Wavetable synthesis

The term 'wavetable synthesis' alludes to a relatively broad range of synthesis methods largely used by commercial synthesisers, ranging from MIDI-keyboards to computer sound cards. Also, expressions such as *vector synthesis* and *linear arithmetic* have been brought into play by the synthesiser industry to refer to slightly different approaches to wavetable synthesis. There are at least four major distinct mechanisms that are commonly employed to implement wavetable synthesis: single wavecycle, multiple wavecycle, sampling and crossfading.

In fact, most synthesis languages and systems, especially those derived from the Music N series, function using wavetables. In order to understand this, imagine the programming of a sinewave oscillator. There are two ways to program such oscillators on a computer. One is to employ a sine function to calculate the samples one by one and output each of them immediately after their calculation. The other method is to let the machine calculate only one cycle of the wave and store the samples in its memory. In this case, the sound is produced by repeatedly scanning the samples as many times as necessary to forge the duration of the sound (Figure 3.31). The memory where the samples are stored is technically called *wavetable* or *lookup table*; hence the scope for the ambiguity we commonly encounter with this term.

Figure 3.31 The samples for a waveform are stored on a wavetable and the sound is produced by repeatedly scanning the samples

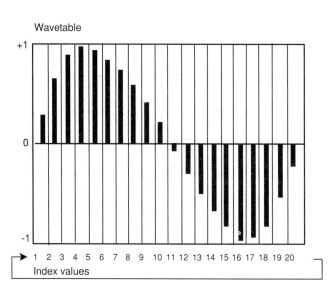

Figure 3.32 The single
wavecycle approach to
wavetable synthesis

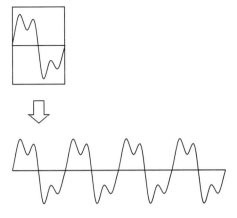

Figure 3.32 The single
wavecycle approach to
wavetable synthesis

Figure 3.33 Single
wavecycle concatenating
between straight and reverse
playback

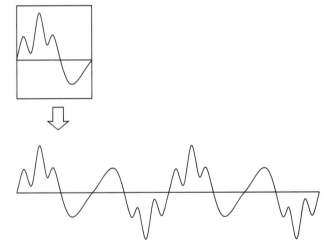

3.6.1 *Single wavecycle*

The single wavecycle approach works with wavetables containing only one cycle of sounds. This is normally the case for oscillator units found in software synthesis systems; e.g. pcmusic *osc*. The result is a repetitive static sound without any variance in time (Figure 3.32). Single wavecycle-based synthesisers normally provide a bank of waveforms such as sine, triangle and sawtooth, and in some systems these waveforms can be mixed together for playback. More complex set-ups also allow for the selection of different methods for playing back the cycle; for instance, by concatenating between straight and reverse playback or even by concatenating different types of waveforms (Figure 3.33).

In order to introduce some variance over time, wavecycle oscillators should be used in conjunction with signal modifiers such as envelopes and low-frequency oscillators (LFO).

3.6.2 Multiple wavecycle

The multiple wavecycle approach is similar to that of the single wavecycle approach. The main differences are that a wavetable may hold more than one cycle of a sound and the samples are usually from recordings of 'natural' instruments, as opposed to synthetically generated waveforms. This is the approach adopted by the manufacturers of most computer sound cards. These cards usually have a bank of recorded timbres in read-only memory (ROM) plus space in random access memory (RAM) to add a few others, either sampled by the user or provided by a third party (e.g. on CD or from the Internet). Additional effects and simple sound-processing tools are often provided to allow some degree of user customisation. For example, an envelope can be redrawn or a low-pass filter can be used to dump higher partials. Other sophisticated mechanisms, such as using a sequence of partial wavecycles to form a complete sound, are commonly associated with the multiple wavecycle approach; for example, when the attack, decay, sustain and release portions of the sound are taken from different sources (Figure 3.34).

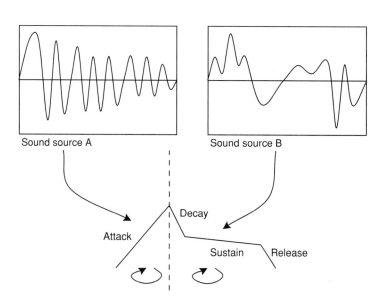

Figure 3.34 Multiple wavecycle using two different sound sources for different portions of the sound

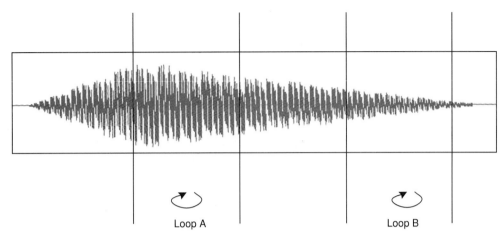

Figure 3.35 Sampling with two internal loopings

3.6.3 Sampling

In some cases it may be advantageous to use longer wavetables. The difference between multiple wavecycling and sampling is that the latter generally uses longer wavetables. Even though the multiple wavecycle approach works with more than one cycle, the size of the wavetables is relatively short and may need several loops to produce even a note of short duration. One advantage of sampling over multiple wavecycling is that longer wavetables allow for the use of pointers within a sample in order to define internal loopings (Figure 3.35). Due to this flexibility, sampling is normally used for creating sonorities and effects that would not be possible to create acoustically.

Figure 3.36 Crossfading works by gradually changing the sample outputs from one wavetable to the sample outputs of another

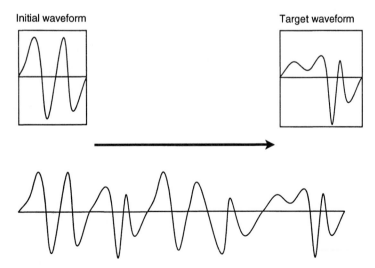

3.6.4 Crossfading

Crossfading is an approach which combines elements of multiple wavecycling and sampling. This works by gradually changing the sample outputs from one wavetable by the sample outputs of another in order to produce mutating sounds (Figure 3.36). Crossfading is a simple way of creating *sound morphing* effects, but it does not produce good results all the time. Some architectures let musicians specify more than only two sources to crossfade, and more sophisticated set-ups allow for the use of manual controllers, such as a joystick, to drive the crossfading. Variants of the crossfading approach use alternatives techniques (e.g. sample interpolation) in order to make the transition smoother.

3.6.5 Wavetable synthesis examples

On the accompanying CD-ROM Nicky Hind's CLM tutorial (in folder *Clm*) introduces basic techniques for processing digitally sampled sounds in CLM. Virtual Waves (in folder *Virwaves*) provides a collection of tools for making instruments that use sampled sounds as their main source signal. There are a number of instruments of this type in folder *Synthvir*; for example, *Android1.syn*, *Atomic.syn*, *Cut-up2.syn*, *Cyborg.syn* and *Zozozong.syn*. These instruments perform specific operations on the sampled sound in order to produce the desired effect; for instance, the instrument *Android.syn* amplitude modulates the signal in order to create extra sidebands in the spectrum of the sampled sound. In this same lot of examples, *Crossfad.syn* is an

Figure 3.37 The *Spectral Sketch Pad* module of Virtual Waves allows for the creation of sounds graphically on a sketch pad area

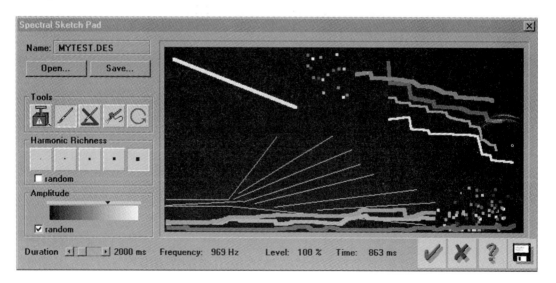

admirable case of crossfading synthesis where the sound of a cello changes to a vocal sound. But the most genuine example of wavetable synthesis is *Drawing1.syn*, which features a unique Virtual Waves synthesis module: the *Spectral Sketch Pad* module (Figure 3.37). This module allows for the creation of sounds graphically on a sketch pad area where the horizontal axis corresponds to time control and the vertical axis to frequency control. There are three sketching tools available:

- The paintbrush: for drawing freehand lines
- The straight line pencil: for drawing straight lines
- The airbrush: for creating patches of spots, or 'grains' of sound

4 Physical modelling techniques

Much literature and advertising has been fairly misleading by inconsistently using terms such as *virtual acoustics* and *waveguides* as synonyms for physical modelling. Physical modelling (PM) synthesis is by no means a single synthesis technique. It is rather a class of different techniques that share some fundamental principles. However, because there are controversies as to what these fundamentals are, there are no definitive criteria to establish whether or not a certain technique qualifies as PM. This chapter deliberately expands the scope of the physical modelling ideal in order to include a number of synthesis techniques that some might not consider as such; for example, the *recirculating wavetable* techniques.

Despite all controversies, it is generally agreed that PM works by emulating the functioning of acoustic musical instruments; the key issue of PM is the emulation of acoustic sound generators rather than the sounds themselves. PM emulation can be achieved in a variety of ways. The classic method starts from a system of equations describing the acoustic behaviour of an instrument when it is played. In principle this method is appealing, but it is not yet practical for musicians: the equations are too complex to design and the computer technology required to process these equations is still too expensive for it to be accessible. A variety of alternatives has been proposed, however, to alleviate this burden; for example, the modularisation of equations into functional blocks. In this case, one builds an instrument by assembling these blocks.

The rationale for the desirability of PM is that it can produce a convincing emulation of most acoustical instruments, including the production of expressive sound attributes such as breathing, sliding fingers, etc. Moreover, it allows for the design of virtual instruments that would otherwise be impossible to build; for example, a morphing didgeridoo-like instrument that shrinks to the size of a flute, changing its material from wood to metal, during the course of a melody. Note, however, that not all PM techniques are capable of fulfilling all aspects of our imagination. Some techniques may excel for certain modelling aspects but may completely fail for others.

The physical modelling ideal is not new. It had been used to study the behaviour of musical instruments long before electronic sound synthesis could be envisaged. Some would even argue that physical modelling dates back to the early days of Western science when Pythagoras designed the monochord: a model to study music theory.

The concepts behind PM synthesis as such can be traced to nineteenth-century scientific treatises on the nature of vibrating objects; for example, *The Theory of Sound*, first published in 1877, by J. W. S. Rayleigh, a Fellow of Trinity College in Cambridge.

Treatises on vibrating systems often describe the behaviour of musical instruments with the aid of mechanico-acoustical models. It was, however, the emergence of electrical models that fostered the development of PM synthesis. Electrical circuits can be described by equations that are analogous to the equations used to describe mechanical and acoustical vibrating systems. Electrical models were widely used to study the emulation of the acoustic behaviour of musical instruments because electrical engineering had developed powerful tools for circuitry problem solving that could be applied to the solution of vibration problems in acoustics.

It is known that various electrical models of musical instruments were created as early as the 1930s. At the beginning of the 1950s, for example, H. F. Olson, of RCA Laboratories in Princeton, USA, published a book entitled *Musical Engineering*, containing various examples of electrical modelling of musical instruments. In this book, Olson illustrates how the study of mechanico-acoustical systems is facilitated by the introduction of elements analogous to the elements in an electrical system (Figure 4.1).

Progress in turning these models into usable prototypes for music making was slow until computer technology became available. Computer modelling of vibrating systems is normally

(a) Helmholtz resonator

(b) Electrical model

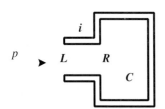

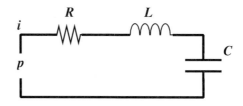

Figure 4.1 Electrical circuitry (b) may be used to model mechanico–acoustical systems (a)

based upon the mechanical and not upon the electrical paradigm, because musicians can more easily monitor the simulation with the aid of on-screen displays of the 'mechanics' of the system.

An accurate physical model of a musical instrument requires a great amount of computation, no matter what modelling metaphor is used. Various simplified PM synthesis alternatives have been designed to bypass the complex mathematical nature of the classic physical modelling of acoustic systems and the heavy computational burden they impose. Despite controversies as to whether or not 'simplified' PM alternatives are truly physical models, it must be said that most of them can cheat the human ear incredibly well. The waveguide technique, developed at Stanford University in California, USA, is one example of these successful alternatives (Smith, 1992).

Simplified physical modelling methods combined with state-of-the-art chip technology enabled the industry to release the first samples of commercial PM synthesisers at the beginning of the 1990s: Bontempi-Farfisa's MARS in 1992 followed by Yamaha's VL1 in 1993.

4.1 Approaches to physical modelling

4.1.1 Classic approach

The classic PM approach emulates the behaviour of vibrating media using a network of interconnected mechanical units, called *mass* and *spring* (Figure 4.2). On a computer, this network is implemented as a set of differential equations, whose solution describes the waves produced by the model in operation. Sound samples result from the computation of these equations.

This approach has the ability to capture two essential physical properties of vibrating media: *mass density* and *elasticity*. A single string, for example, can be modelled as a series of masses

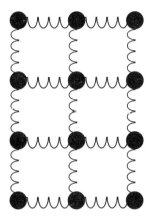

Figure 4.2 The behaviour of vibrating media may be emulated using a network of interconnected mechanical units

connected by springs (Figure 4.3). The density of the string model is determined by the amount of mass units and its elasticity by the strength of the springs. To begin with, the model is in a state of equilibrium. If force is applied to a certain point, then the model is displaced from its position of equilibrium (e.g. equivalent to plucking the string). Disturbed mass units propagate the disturbance and generate wave motion by forcing adjacent masses to move away from their point of equilibrium (Figure 4.4). The compromise between the mass density and the elasticity of the model defines the speed of the propagation and the amount of resistance of the model to disturbance and the time it takes to restore its equilibrium.

Surfaces (e.g. the skin of a drum) and volumes (e.g. the body of a guitar) can be modelled as a network of masses connected by more than one spring (Figure 4.2).

The PhyMod program by Joerg Stelkens (in folder *Various/Phymod* on the CD-ROM) is a good example of a classic approach to PM on affordable PC-compatible platforms.

4.1.2 Functional approach

Most musical instruments can be modelled as a resonating chamber stimulated by acoustic waveforms with certain spectral and temporal properties. The functional approach to PM is based upon the principle that the behaviour of an instrument is determined by two main components: *source* and *resonator*. The role of the source component is to produce a raw signal that will be

Figure 4.3 A string can be modelled as a series of masses connected by springs

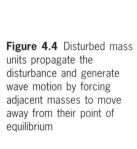

Figure 4.4 Disturbed mass units propagate the disturbance and generate wave motion by forcing adjacent masses to move away from their point of equilibrium

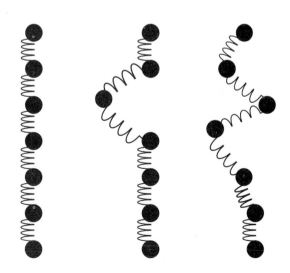

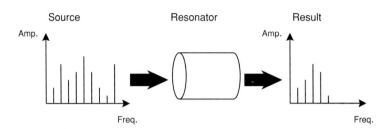

Figure 4.5 A variety of
spectra can be obtained by
varying the acoustic
properties of the resonator

shaped by the resonator (Figure 4.5). A variety of spectra can be obtained by varying the acoustic properties of the resonator. For example, air blown past a cylindrical tube will produce a different sound quality, according to the dimensions of the tube.

From a signal processing point of view, the resonator acts as a filter (or bank of filters) applied to a non-linear (i.e. noisy) source signal. Emulation of various sound spectra is achieved by combining the production of suitable source signals with the specification of appropriate coefficients for the filters.

The interaction between the source and the resonator components falls into two categories: *feedforward* and *feedback* (also known as decoupled and coupled, respectively). With feedforward, a source signal is simply injected into a resonator. Conversely, with feedback, the resulting signal from the resonator is fed back to the source (Figure 4.6). The latter category often produces more convincing results than the former, because it is the interaction between the source and the resonator in most acoustic instruments that creates the subtle effects we hear in performance. For example, the frequency of the vibrating reed of a saxophone is strongly influenced by acoustic feedback from the resonating bore (i.e. 'tube') of the instrument, after being stimulated initially by a blast of air from the mouth.

Figure 4.6 The feedback
interaction between the
source and the resonator is
an important aspect of the
functional approach to PM

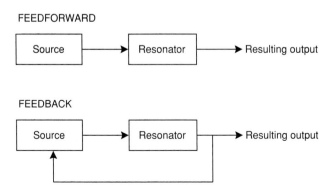

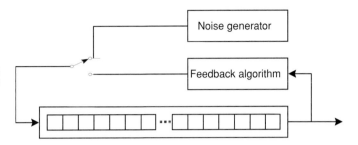

Figure 4.7 The recirculating wavetable approach to PM. As samples are output from the right side of the array, they are processed according to a certain algorithm and the results are fed back in the left side

4.1.3 Recirculating wavetable approach

The recirculating wavetable approach to PM uses a time-varying lookup table to simulate the behaviour of a vibrating medium. The basic functioning of this method starts with a lookup table of a fixed length, filled with random samples. In this case, the table functions as a queue of sample values, rather than as a fixed array, as in the case of a simple oscillator. As samples are output from the right side of the array, they are processed according to a certain algorithm, and the result is fed back in the left side (Figure 4.7). The algorithm for processing the samples defines the nature of the simulation. For example, the averaging of the current output sample with the one preceding it in the array functions as a type of low-pass filter.

Although the recirculating wavetable method does not bear strong physical resemblance to the medium being modelled, its functioning does resemble the way in which sounds produced by most acoustic instruments evolve: they converge from a highly disorganised distribution of partials (characteristic of the initial noise components of the attack of a note) to oscillatory patterns (characteristic of the sustained part of a note).

4.2 Implementation techniques

4.2.1 Modal synthesis

The modal synthesis technique is inspired by the mass–spring paradigm. It starts from the principle that a sound-producing object can be represented as a collection of vibrating substructures. The number of substructures is usually very small in comparison with those in the classic mass–spring approach. Examples of substructures here include violin bridges and bodies, acoustic tubes, bells, drum heads, etc.

Modal synthesis partitions sound-producing mechanisms into vibrating substructures. Each substructure is characterised as a set of modal data consisting of:

- The frequencies and damping coefficients of the substructure's resonating modes
- A set of coordinates representing the shape of the vibrating mode

The Mosaic system, developed by Jean-Marie Adrien at the University of Paris VI, is a typical example of a tool for implementing modal synthesis (Adrien, 1991). It provides a number of ready-made substructures such as strings, air columns, metal plates and membranes, as well as substructures for the simulation of stimulation such as bowing and hammering. Instruments are thus programmed by networking these substructures. In Mosaic terms, substructures are called *objects* and the interactions between objects are referred to as *connections*. A connection between two objects also acts as an interface between the user and the instrument; for example, the connection between a bow and string provides the means for user-customisation of the parameters that drive the interaction such as pressure, speed, etc.

From a musician's point of view, besides reduction of mathematical complexity, the modal approach has the advantage over the mass–spring paradigm of modularity of the substructures. The instrument designer can add or subtract substructures on a network to create time-varying synthesis effects (such as expanding or shrinking the size of an instrument) and/or timbral hybrids by combining substructures from different instruments.

4.2.2 Subtractive synthesis

Subtractive synthesis embodies the functional approach to PM; it models an instrument as a resonating chamber that sculpts a tone out of a raw source signal (Figure 4.5). The implementation of a subtractive synthesiser generally employs a signal generator with a rich spectrum, to act as a source, and a bank of filters to simulate a resonating chamber. This technique is referred to as subtractive, because the filters alter the spectrum of the source signal by subtracting unwanted partials of its spectrum while favouring the resonation of others.

Two types of signal generators are commonly used as source for subtractive synthesis, because of the richness (i.e. in terms of variety of partials) of their output: *noise* and *pulse* generators. Whilst the noise generator produces a large number of random partials within a broad bandwidth, the pulse generator produces a periodic waveform, at a specific frequency, with strong energy in its harmonics. The spectrum of the waveform of the pulse is determined by the ratio of the pulse width to the period of the signal; the smaller the ratio, the narrower the pulse and therefore the higher the energy of the high-frequency partials.

Subtractive synthesis has been used successfully to model percussion-like instruments and the human voice mechanism. The main limitation of this technique is that non-linear interactions between the source and the resonator are ignored, thus imposing difficulties for controlling fine sound characteristics of acoustic instruments.

Brief introduction to filtering

In general, a filter is any device that performs some sort of transformation in the spectrum of a signal. For simplicity, however, in this section we refer only to filters which cut off or favour the resonance of specific components of the spectrum. In this case, there are four types of filters, namely: low-pass (LPF), high-pass (HPF), band-pass (BPF) and band-reject (BRF).

The basic building block of the subtractive synthesiser is the BPF, also known as the resonator. BPF rejects both high and low frequencies with a passband in between. Two parameters are used to specify the characteristics of a BPF: passband centre frequency (represented as f_c) and resonance bandwidth (represented as bw). The bw parameter comprises the difference between the upper (represented as f_u) and lower (represented as f_l) cut-off frequencies (Figure 4.8).

The BRF amplitude response is the inverse of a BPF. It attenuates a single band of frequencies and discounts all others. Like a BPF, it is characterised by a central frequency and a bandwidth; but another important parameter is the amount of attenuation in the centre of the stopband.

Figure 4.8 The basic building block of subtractive synthesis is the band-pass filter (BPF)

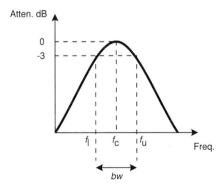

Figure 4.9 The low-pass filter (LPF) reduces the amplitude of spectral components above the cut-off frequency

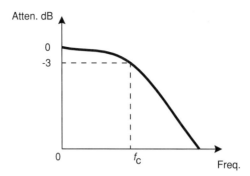

An LPF permits frequencies below the point called the cut-off frequency to pass with little change. However, it reduces the amplitude of spectral components above the cut-off frequency (Figure 4.9). Conversely, an HPF has a passband above the cut-off frequency where signals are passed and a stopband below the cut-off frequency, where the signals are attenuated. There is always a smooth transition between passband and stopband. It is often defined as that frequency at which the power transmitted by the filter drops to one half (about –3 dB) of the maximum power transmitted in the passband.

Under special conditions a BPF may also be used as an LPF or as an HPF. An LPF can be simulated by setting the BPF's centre frequency to zero. The resulting cut-off frequency would intuitively be one-half of the *bw*. As a consequence of this simulation, however, the cut-off frequency of the resulting low-pass is 70.7 per cent of the specified bandwidth, and not 50 per cent. For example, if the desired LPF cut-off frequency is to be 500 Hz, the bandwidth value for the BPF must be 1 kHz multiplied by 1.414 (that is, 1000 × 1.414). This is because the BPF is a two-pole filter; at its cut-off frequency (when output is 50 per cent) the output power of a true LPF of the same cut-off would be in fact 70.7 per cent.

An HPF can be made from a BPF by setting its centre frequency equal to the Nyquist frequency; i.e. the maximum frequency value produced by the system. HPFs made from BPFs suffer from the same approximation problems that affect LPFs.

The rate at which the attenuation increases is known as the slope of the filter – or *rolloff*. The rolloff is usually expressed as attenuation per unit interval, such as 6 dB per octave. In the stopband of an LPF with a 6 dB/octave slope, for example, every time the frequency doubles, the amount of attenuation increases by 6 dB. The slope of attenuation is determined by the order of the filter.

Figure 4.10 A subtractive synthesiser usually requires a combination of interconnected filters

Cascade filter composition

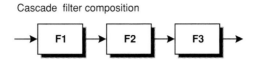

Parallel filter composition

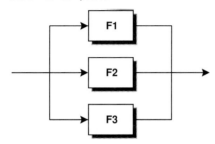

Order is a mathematical measure of the complexity of a filter; in a digital filter, it is proportional to the number of calculations performed on each sample.

Filter composition

The resonator component of a subtractive synthesiser usually requires a composition of interconnected filters in order to produce the desired spectrum. There are two basic combinations for filter composition: *parallel connection* and *serial connection* (also known as *cascade connection*) (Figure 4.10).

In a serial connection, filters are connected like the links of a chain. The output of one filter feeds the input of the next, and so on. The output of the last filter is therefore the output of the entire chain. Much care must be taken when composing filters with different passband centre frequencies in series. In serial connection, the specification of one filter's passband does not guarantee that there will be significant energy transmitted in that passband. If any of the previous elements of the cascade have had significant attenuation in that frequency range, then the response will be affected.

With parallel connection, the signal to be filtered is simultaneously applied to all filter inputs. The outcome is the addition of the frequency responses of all filters, resulting in the resonation of any frequency found within the passband of the individual filters. Each filter of the parallel connection is usually preceded by an amplitude control, which determines the relative amplitudes of the partials of the resulting spectrum.

Figure 4.11 A simple
subtractive synthesis example
from GENERATOR

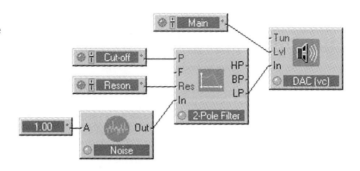

Subtractive synthesis examples

The effect of the low-pass filter (LPF) on noise is illustrated by *Subtract1.ens*, in folder *Ensembs* in the GENERATOR materials (Figure 4.11). The fader named as *Cut-off* changes the cut-off frequency above which the amplitude of the partials will the attenuated.

An example of a subtractive synthesis instrument implemented using Virtual Waves is also available on the CD-ROM: *Subtract.syn*. The signal from a pulse generator is fed onto four second-order band-pass filters in parallel and the output from each filter is amplitude controlled by an envelope. The pulse generator is frequency modulated by a noise generator in order to create a harsh effect at the beginning of the sound.

An in-depth study on the design of a subtractive synthesiser in pcmusic is given in Chapter 7 and a few examples are available on the CD-ROM in folder *Vocsim*, in the pcmusic materials.

4.2.3 Waveguide filtering

The waveguide filtering technique is a type of functional approach to PM. A waveguide filter comprises a variety of signal processing units, mostly delay lines and LPFs. Unlike subtractive synthesis, waveguide filtering is based on the *feedback* interaction between the source and the filter component. Waveguide filtering is therefore able to model resonating media other than acoustic chambers, such as strings.

The best way to visualise the functioning of this technique is to imagine what happens when the string of a monochord is struck at a specific point: two waves travel in opposite directions and when they reach the bridges, some of their energy is absorbed and some is reflected back the point of impact, causing resonances and interferences (Figure 4.12).

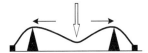

Figure 4.12 When the string of a monochord is struck at a specific point, two waves travel in opposite directions. When they reach the bridges, some of their energy is absorbed, and some is reflected back to the point of impact, causing resonances and interferences

A generic waveguide filter instrument is illustrated in Figure 4.13. A source signal – usually noise – is input into a bi-directional delay line and travels until it reaches a filter (Filter A). This filter passes some of the signal's energy through and bounces some energy back; it models the effect of a 'scattering junction', such as a hole in a cylindrical tube or a finger pressing a string. The other filter (filter B) at the end of the chain models the output radiation type; for example, the bell of a clarinet.

The input source signal for the delay line and the type of modification applied by the waveguide chain play a key role in determining the characteristics of the instrument. Waves running forwards and backwards along the chain cause resonances and interferences at frequencies related to its dimensions. When the waveguide network is symmetrical in all directions, the sound it produces when stimulated will tend to be harmonic. If the waveguide twists, changes size, or intersects with another waveguide, then its resonant pattern is changed. This synthesis technique is largely compatible with most available sound synthesis programing languages; standard unit generators can be used as the building blocks of waveguide networks.

Figure 4.13 A generic waveguide filter instrument

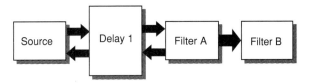

Waveguide filtering examples

Figure 4.14 illustrates a simple simulation of a pipe instrument implemented in Audio Architect; try *Pipe1.mae* in folder *Aamets* in the Audio Architect materials on the accompanying CD-ROM. The acoustic counterpart consists of a pipe open on one side and some kind of blowing mechanism on the other. When someone blows the instrument, a signal travels down the pipe. When this signal reaches the other end it spreads out into the air creating disturbances that are perceived as sound. The laws of physics teach us that the air offers some resistance to disturbance thus making part of the outcoming energy bounce back inside the pipe towards the blowing source. The feedback portion then travels back towards the open end, and the loop continues until the signal fades away. Considering that the signal travels inside the pipe at one foot per second, it would take 8 milliseconds to

Figure 4.14 A simple simulation of a pipe instrument in Audio Architect

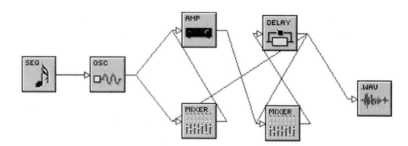

reach the open end of the pipe and to bounce back to the blowing source. This phenomenon is simulated in *Pipe1.mae* by an 8-millisecond delay mechanism (the DELAY unit) that feeds back the delayed signal onto itself.

The blowing mechanism is simulated here by an oscillator (the OSC unit). We asssume that the lips of the musician produce a periodic signal onto the pipe and in order to create this signal the lips generate rapid streams of pressure bursts. When the signal bounces back, the lips could therefore be closed or open. If they are open the sound that is coming back will collide with the newly produced burst thus producing a different reflection as if the lips were closed; the signals may even cancel each other. This is simulated by introducing an amplitude modulation arrangement (the AMP unit) before the signal is fed into the delay. Notice that the output from the oscillator plays a twofold role in *Pipe1.mae*: it acts as a source signal and as a modulator. The idea here is that the signal travels down the pipe and each time it bounces back it is amplitude modulated by the rapid opening and closing of the lips. By feeding the output of the delay back into itself and back to the modulator, one can roughly simulate the behaviour of the pipe and the action of the lips. More advanced examples of waveguide programming are included in the CLM tutorial on the accompanying CD-ROM.

4.2.4 Karplus–Strong technique

The Karplus–Strong technique is a classic example of the recirculating wavetable approach to PM. The algorithm, devised at Stanford University by Kevin Karplus and Alex Strong, averages the current output sample of a delay line with the preceding one, and feeds the result back at the end of the delay queue (Karplus and Strong, 1983). Note that in this case the delay line and the wavetable (or lookup table) are the same thing.

This technique produces convincing simulations of the sound of a plucked string. The signal bursts with a loud, bright percussive-like quality, then it darkens and turns into a simpler sinusoidal type of sound. The decay of the sound is normally dependent upon the size of the wavetable, but it is possible to control the decay length by adding time-stretching mechanisms to the original algorithm.

Indeed, several variations on the original Karplus–Strong technique have been developed in order to produce effects other than the plucked string. For example, simulations of drum-like timbres can be achieved by inverting the sign of a few output samples, according to a probability factor.

Some software packages for sound synthesis programing provide a special unit for the implementation of the Karplus–Strong technique, which normally needs five parameters to function:

- Amplitude
- Frequency
- Size of the recirculating lookup table
- The nature of the initialisation
- The type of feedback method

The size of the recirculating lookup table is specified in Hertz and is usually set equal to the frequency value. The lookup table can be initialised by either a random sequence of samples or a user-defined function. As for the feedback method, the simple averaging method described above is the most commonly used.

Karplus–Strong synthesis examples

Virtual Waves (in folder *Virwaves*) provides a module which encapsulates the Karplus–Strong algorithm. On the accompanying CD-ROM, the example *String.syn* (in folder *Synthvir*) illustrates a simple instrument that produces plucked string-like sounds using Virtual Wave's *Karplus–Strong Algorithm*.

A more flexible implementation of a Karplus–Strong instrument is illustrated by Nicky Hind in his CLM tutorial (in folder *Clm*). Also, a couple of examples in pcmusic are available in folder *Pluck*, in the pcmusic materials on the CD-ROM.

4.2.5 Cellular automata lookup table

The cellular automata lookup table is a type of recirculating wavetable technique whose functioning resembles the Karplus–Strong technique introduced above. The difference is that the cellular automata lookup table technique uses cellular

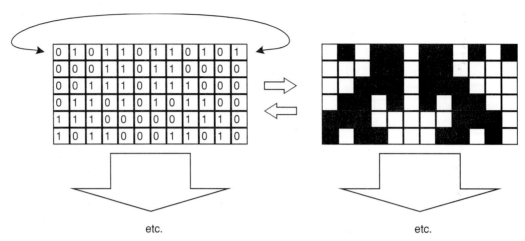

etc.   etc.

Figure 4.15 A cellular automata is implemented as a regular array of variables called cells. Each cell may assume values from an infinite set of integers and each value is normally associated with a colour, in this case 0 = white and 1 = black

automata to calculate the new sample values of the delay line, rather than the averaging method.

Cellular automata are computer modelling techniques widely used to model systems in which space and time are discrete and quantities take on a finite set of discrete values (Wolfram, 1986). Physical systems containing many discrete elements with local interaction can be modelled using cellular automata. In principle, any physical system described by differential equations may be modelled as cellular automata.

Cellular automata are implemented as a regular array of variables called cells. Each cell may assume values from a finite set of integers. Each value is normally associated with a colour. The functioning of a cellular automata is displayed on the computer screen as a sequence of changing patterns of tiny coloured cells, according to the tick of an imaginary clock, like an animated film. At each tick of the clock, the values of all cells change simultaneously according to a set of rules that takes into account the values of their neighbourhood (Figure 4.15). Although cellular automata normally have two or three dimensions, the recirculating lookup table technique uses only one-dimensional cellular automata, and the cells can assume only two values: on (represented by 1) or off (represented by 0).

An example set of rules for the functioning of such types of automata can be defined as follows:

1	if	111	then 0
2	if	110	then 1
3	if	101	then 0
4	if	100	then 1

5	if	011	then 1
6	if	010	then 0
7	if	001	then 1
8	if	000	then 0

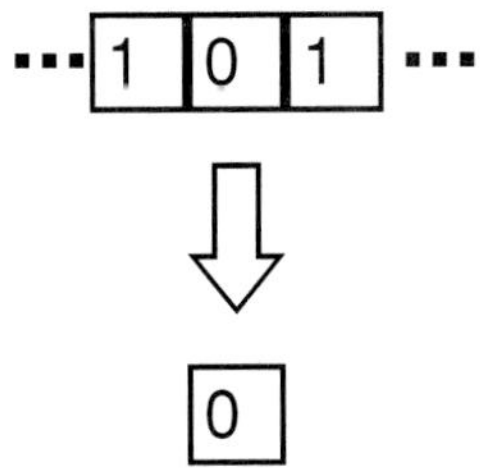

Figure 4.16 Rule 3 states that if a cell is in state 0 and the state of its two neighbours is 1, then this cell remains in state 0 in the next tick of the CA clock

The eight possible states of the three adjacent cells (i.e. 3 adjacent cells = 2^3 rules) define the state of the central cell on the next step of the clock (Figure 4.16).

The synthesis method works by considering the array of cells of the one-dimensional cellular automata as a lookup table; each cell of the array corresponds to a sample. The states of every cell are updated at the rate of the cellular automata clock and these values are then heard by piping the array to the digital-to-analog converter (DAC).

Interesting sounds may be achieved using this technique, but the specification of suitable cellular automata rules can be slightly difficult. However, once the electronic musician has been gripped by the fascinating possibilities of this technique, an intuitive framework for the specification of cellular automata rules should naturally emerge; for example, a rule of a 'growing' type of behaviour (i.e. more and more cells are activated in time) invariably produce sounds whose spectral complexity increases with time.

The LASy program on the CD-ROM was especially designed for experimentation with cellular automata lookup table (see also Chapter 2).

5 Time-modelling techniques

Philosophical considerations aside, time is one of the most important parameters of sound. The great majority of acoustic, synthesis and music descriptors depend upon some timing factor; e.g. wavecycle, frequency, samples, sampling rate, rhythm, etc. It is therefore not surprising that some synthesis techniques approach the problem from a time-domain perspective.

In a lecture entitled *Four Criteria of Electronic Music*, given in London in 1971, the composer Karlheinz Stockhausen introduced some inspiring time-domain sound-processing methods that played an important role in the development of computer tools for time modelling (Maconie, 1989). The general idea of Stockhausen was that if two different pieces of music, such as a symphonic piece by Villa Lobos and a Scottish folk tune, are accelerated until they last only a couple of seconds each, the results would be two different sounds with very distinct timbres. Conversely, if two different short sounds are stretched out in time, then the result would be two different sound streams whose forms reveal the inner time structures of the original sounds. Based upon this idea, in the 1950s Stockhausen devised a very peculiar way to synthesise sounds, which is perhaps the first case of a time-modelling technique. He recorded individual pulses on tape, from an electronic generator. Then he cut the tape and spliced the parts together so that the pulses could form a particular rhythm. Next he made a tape loop of this rhythm and increased the speed until he could hear a tone. Various tones could be produced by varying the speed of the loop. Different rhythms on tape produced different timbres; the components of the rhythmic sequence determined the spectrum according to their individual cycle on the tape loop.

The most interesting aspect of this method, from a composer's point of view, is that it encourages one to work with rhythm and pitch within a unified time domain. It should be noted, however, that the transition from rhythm to pitch is not perceived precisely. The human ear can hear a distinct rhythm up to approximately 10 cycles per second but a distinct pitch does not emerge until approximately 16 cycles per seconds. It is therefore not by chance that the categorical differentiation between rhythm and pitch has firmly remained throughout the history of Western music.

On the whole, Stockhausen's imgemious work gave rise to a number of approaches to current time modelling synthesis. The postulate here is that sounds may be specified entirely by the variations of cyclic patterns in the time domain. In this case, musically interesting sounds may be synthesised by forging the transformations of its waveform as they develop in time. In a broader sense, time modelling synthesis thus uses techniques that control wave transformations. Time modelling is particularly suited for the synthesis of highly dynamic sounds in which most of its properties vary during their course.

5.1 Granular synthesis

Granular synthesis works by generating a rapid succession of very short sound bursts called *grains* (e.g. 35 milliseconds long) that together form larger sound events. The results generally exhibit a great sense of movement and sound flow. This synthesis technique can be metaphorically compared with the functioning of motion pictures in which an impression of continuous movement is produced by displaying a sequence of slightly different images at a rate above the scanning capability of the eye.

The composer Iannis Xenakis is commonly cited as one of the mentors of granular synthesis. In the 1950s, Xenakis developed important theoretical writings where he laid down the principles of the technique (Xenakis, 1971). The first computer-based granular synthesis system did not appear, however, until Curtis Roads (1988) and Barry Truax (1988) began systematically to investigate the potentials of the technique in the 1970s, in the USA and Canada respectively.

5.1.1 The concept of sound grains

In order to gain a better understanding of granular synthesis it is necessary to comprehend the concept of *sound grains* and their relationship to our auditory capacity. This notion is largely based upon a sound representation method published in 1947, in a paper by the British physicist Dennis Gabor (1947). Until then the representation of the inner structure of sounds was chiefly based upon Fourier's analysis technique and Helmholtz's musical acoustics (Helmholtz, 1885).

In the eighteenth century, Fourier proposed that complex vibrations could be analysed as a set of parallel sinusoidal frequencies, separated by a fixed integer ratio; for example, 1x, 2x, 3x, etc., where x is the lowest frequency of the set. Helmholtz then developed Fourier's analysis for the realm of musical acoustics. He proposed that our auditory system identifies musical timbres by decomposing their spectrum into sinusoidal components, called *harmonic series*. In other words, Helmholtz proposed that the ear naturally performs Fourier's analysis to distinguish between different sounds. The differences are perceived because the loudness of the individual components of the harmonic series differs from timbre to timbre.

The representation of such analysis is called *frequency-domain representation* (Figure 5.1). The main limitation of this representation is that it is timeless. That is, it represents the spectral components as infinite, steady sinusoids.

Although now it is taken for granted that the components of a sound spectrum vary substantially during the sound emission, these variations were not considered relevant to be represented until recently. In his paper, Gabor proposed the basis for a representation method which combines frequency-domain and time-domain information. His point of departure was to acknowledge the fact that the ear has a time threshold to discern sound properties. Below this threshold, different sounds are heard as

Figure 5.1 Frequency-domain representation is timeless because it represents the spectrum of a sound as infinite, steady sinusoids

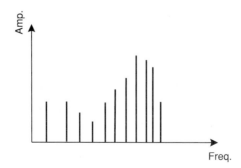

Figure 5.2 Frequency-domain representation snapshots can show how the spectrum of a sound evolves with time

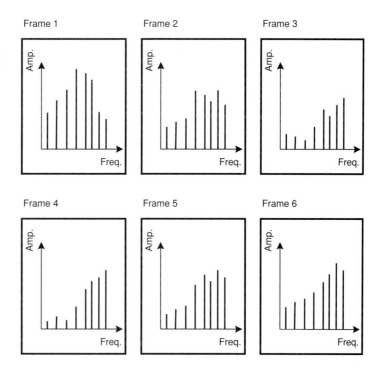

clicks, no matter how different their spectra might be. The length and shape of a wavecycle define frequency and spectrum properties, but the ear needs several cycles to discern these properties. Gabor called this minimum sound length *acoustic quanta* and estimated that it usually falls between 10 and 30 milliseconds, according to the nature of both the sound and the subject.

Gabor's theory fundamentally suggested that a more accurate sound analysis could be obtained by 'slicing' the sound into very short segments, and by applying a Fourier-type of analysis to each of these segments. A sequence of frequency-domain representation snapshots, one for each segment, would then show – like the frames of a film – how the spectrum evolves with time (Figure 5.2). Granular synthesis works by reversing this process; that is, it generates sequences of carefully controlled short sound segments, or *grains*.

5.1.2 Approaches to granular synthesis

As far as the idea of sound grains is concerned, any synthesiser producing rapid sequences of short sounds may be considered as a granular synthesiser. However, there are important issues to consider when designing a granular synthesis instrument.

Figure 5.3 The sequential approach to granular synthesis works by synthesising sequential grain streams

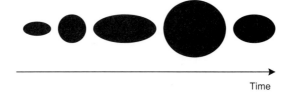

Time

Three general approaches to the technique are presented as follows.

Sequential approach

The sequential approach to granular synthesis works by synthesising sequential grain streams. The length of the grains and the intervals between them are controllable, but the grains must not overlap (Figure 5.3).

Scattering approach

The scattering approach uses more than one generator simultaneously to scatter a fair amount of grains, not necessarily in synchrony, as if they were the 'dots' of a 'sonic spray'. The expression *sound clouds* is usually employed by composers to describe the outcome of the scattering approach (Figure 5.4).

Granular sampling approach

Granular sampling employs a *granulator* mechanism that extracts small portions of a sampled sound and applies an envelope to them. The granulator may produce the grains in a number of ways. The simplest method is to extract only a single grain and

Figure 5.4 The scattering approach to granular synthesis uses more than one generator simultaneously to scatter a fair amount of grains, not necessarily in synchrony

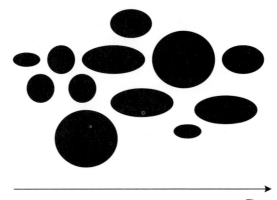

Time

Original sound

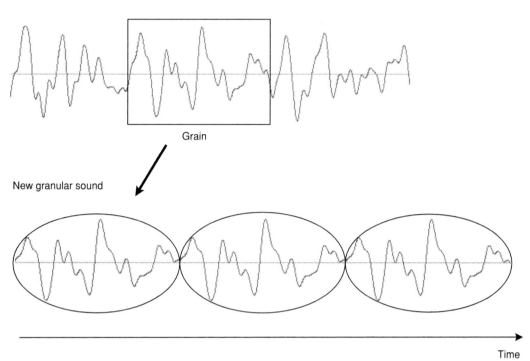

Grain

New granular sound

Time

Figure 5.5 Granular sampling works by replicating a small portion extracted from a sampled sound

replicate it many times (Figure 5.5). More complex methods involve the extraction of grains from various portions of the sample. In this case, the position of the extraction can be either randomly defined or controlled by an algorithm (Figure 5.6). Interesting results can be obtained by extracting grains from more than one sound source (Figure 5.7). Other common operations on 'granulated' sounds include: reversing the order of the grains, time expansion (by changing the time interval between the grains) and time compression (by truncating grains). GranuLab, by the composer Rasmus Enkman, for PC-compatible computers is an example of a program for granular sampling (in folder *Various/Granulab* on the accompanying CD-ROM).

5.1.3 Instrument design

In granular synthesis, sound complexity is achieved by the combination of large amounts of simple grains. The architecture of a granular instrument may therefore be as simple as a single oscillator and a Gaussian-like bell-shaped envelope (Figure 5.8).

Original sound

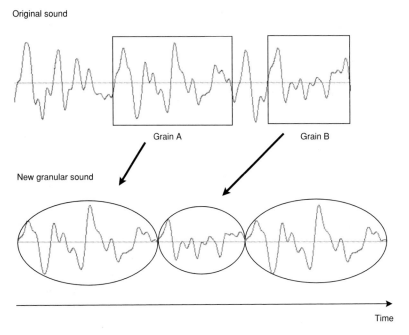

Figure 5.6 A granular sampler may also extract grains from different parts of the sample

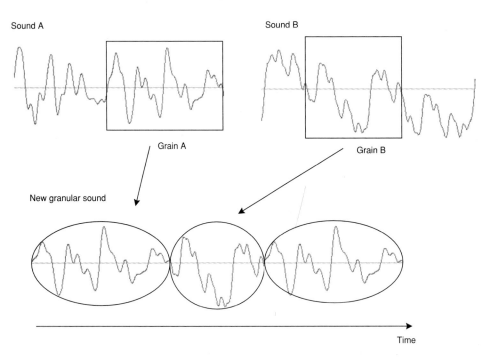

Figure 5.7 Interesting granular sampling effects can be obtained by combining grains extracted from more than one sound source

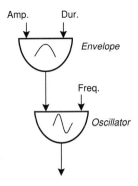

Amp.   Dur.

Envelope

Freq.

Oscillator

Figure 5.8 A granular synthesis instrument may be as simple as a single oscillator and a Gaussian-like bell-shaped envelope

The envelope is of critical importance in granular synthesis because it prevents the glitches that would otherwise be produced by possible phase discontinuities between the grains (Figure 5.9). The shape of the envelope should be carefully chosen because unwanted distortions may be produced by unavoidable 'envelope modulation' (i.e. ring modulation) effects (see Chapter 3). Also, envelopes that use linear segments with angled joints should be used with caution, if at all, as they often produce undesirable spectral components. Sometimes these 'undesirable' effects may be useful in order to add a certain colouration to the resulting sound, but as a rule of thumb they should be avoided.

Despite the simplicity of a granular instrument, its operation requires the control of large sets of parameters. Each grain needs to be specified, but it is impracticable to specify these parameters manually. To begin with, it is very time consuming and difficult to abstract the overall outcome. It would be equivalent to having manually to write the values for thousands of samples in order to synthesise a sinewave. The instrument designer is therefore urged to provide global grain parameters and to devise high-level mechanisms to control them. The global grain parameters commonly found in most granular synthesis programs are:

- Grain envelope shape
- Grain waveform
- Grain frequency
- Grain duration
- Delay time between grains
- Spatial location (e.g. in the stereo field)

High-level mechanisms to control these parameters are normally implemented using one or a combination of the following techniques:

- Trajectory envelopes
- Random values
- Probabilities
- Fractal geometry
- Cellular automata

Figure 5.9 Discontinuities between grains produce undesirable glitches. This may be prevented by using a suitable envelope

In addition to global grain parameters and their control mechanisms, a granular synthesiser should also provide controllers at the level of the whole sound event, such as duration of the event.

On the accompanying CD-ROM, in folder *Chsynth*, there is Chaosynth, a granular synthesis program that uses cellular automata to control the production of the grains (see also Chapter 8). Also, GrainWave for Macintosh computers provides *Grain*, an operator for granular sampling in which a sampled grain is repeatedly triggered according to user-specified frequency settings.

5.2 Resynthesis by fragmentation and growth

The operation of this method is akin to the granular sampling approach to granular synthesis. The postulate of resynthesis by fragmentation and growth is, however, slightly different. Instead of producing huge quantities of short sounds to form larger sound events, this technique basically enlarges a short sound by repeating segments of itself (Figure 5.10).

Resynthesis by fragmentation and growth originates from one of the earliest analog electronic music composition techniques, known as 'tape looping'. Tape looping is a process whereby a short length of recorded tape is cut and both ends are spliced together to form a loop. A repetitive sound pattern is generated by playing back this loop. The music potential of this technique was extensively explored by composers of early electronic music.

Figure 5.10 Resynthesis by fragmentation and growth basically works by enlarging a short sound by repeating segments of itself

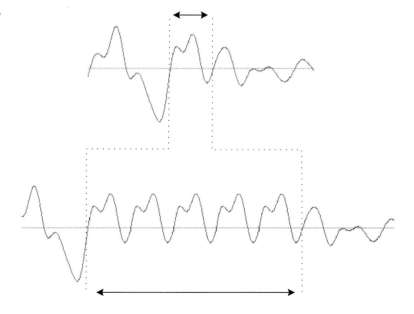

Figure 5.11 Brassage
works by slicing a sound and
then splicing the fragments in
a number of different ways

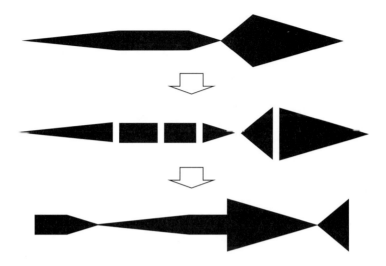

Similar techniques are now available on some MIDI-samplers to simulate sustain on a note that is longer than the sampled sound itself. A number computer programs for segmentation and growth have been developed by the Composer's Desktop Project (CDP) in England (CDP is on the accompanying CD-ROM).

5.2.1 Brassage

Brassage is a technique whereby a sound is fragmented and spliced back in a variety of ways. If the fragments are spliced in the same manner as they were cut, then the original sound will

Figure 5.12 Brassage can extend the duration of a sound by slicing overlapping segments but not overlapping the segments in the resulting sound

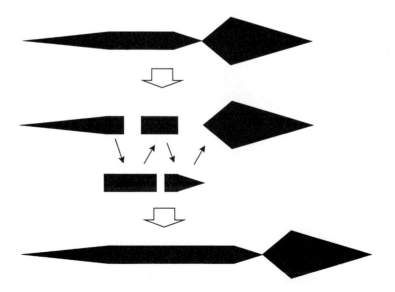

Figure 5.13 Brassage can shorten the duration of a sound by slicing segments with a space between them and then by splicing back the segments without the spaces

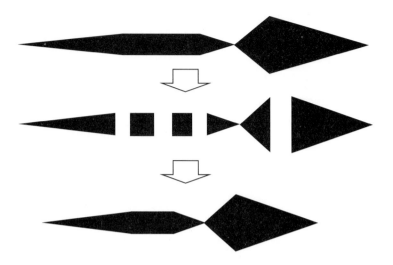

be reproduced. But if, for example, the fragments are shuffled and overlaid, then a completely different, and often bizarre, result may be obtained (Figure 5.11).

The duration of a sound can be extended by cutting overlapping segments from the source but not overlapping them in the resulting sound (Figure 5.12). Conversely, the sound can be shortened by choosing fragments from the source with a space between them and splicing them back without the spaces (Figure 5.13). More sophisticated segmentation may involve the specification of different segment sizes and the variation of either or both the pitch and the amplitude of the fragments. These variations may, of course, be random or driven by a function or envelope.

Figure 5.14 An overlap crossfade window is normally used to avoid discontinuities between the segments in the brassage process

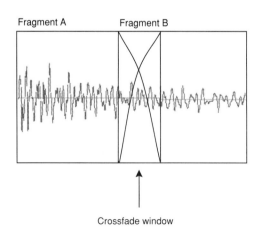

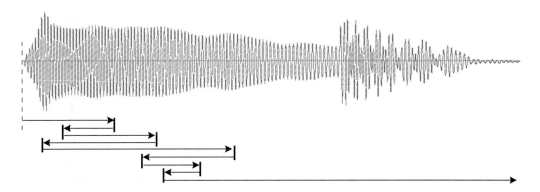

Figure 5.15 The reversal points of a zigzag process can be specified anywhere in the sound and can be of variable length

In order to avoid discontinuities between the fragments, brassage should normally join them together using an overlapping crossfade window (Figure 5.14). If the fragment sizes are very short and uniform, then the crossfading may cause a similar envelope modulation effect as with granular synthesis. In most cases an exponential crossfade curve produces better splicing.

5.2.2 Zigzag

Zigzag is a process in which a sound is played in a zigzag manner. The reversal points can be specified anywhere in the sound and can be of variable length. If the process starts at the beginning of the sound and stops at its end (Figure 5.15), then the result will appear to be artificially extended. The aural effect will depend upon the nature of the original sound. The extension effect will be more convincing if zigzag is applied to highly homogeneous or very 'noisy' sounds.

Interesting results may be obtained with controlled changes in length and position, thus producing 'undulatory' sound effects. As in brassage, zigzag also requires the careful specification of a crossfade window to smooth the continuation of the playback. Again, envelope modulation is bound to occur if the zigzag span is too short and too fast.

An abridged version of CDP's ZIGZAG program is available on the CD-ROM, in folder *Cdp*. Use a Web browser to read the user manual (*Cdpman.htm*) for more information about ZIGZAG.

5.3 Waveset distortion

Waveset distortion involves the transformation of a sound at its wavecycle level; that is, the unit for waveset distortion is a set of samples between two successive zero crossings. For example,

117

a simple sinusoid sound is viewed in this context as a succession of wavecycles, and not as a succession of samples.

As in resynthesis by fragmentation and growth, waveset distortion is also akin to the granular sampling approach to granular synthesis. The main difference here is that radical changes are applied to the individual wavecycles, rather than mere substitution of either the order or location of grains. Also, there is an explicit unit definition to work with here – the wavecycle – as opposite to granular synthesis, where the size of the grain is an implicit subjective unit. Note that in this case the concept of pitch does not always apply. For a simple periodic sound, the greater the number of wavecycles, the higher the pitch, but for a more 'wobbly' sound this may not be the case.

The composer Trevor Wishart has developed a number of waveset distortion programs for the CDP package. There are two types of program in Trevor's kit: programs that work on individual wavecycles and those that work on groups of wavecycles (that is, wavesets). Typical waveset distortion operations are:

- Addition or removal of wavecycles in a waveset
- Modulation of wavecycles or wavesets
- Application of amplitude envelopes to wavecycles
- Replacement of wavesets by other waveforms (e.g. sine, square)
- Permutation of wavecycles

There are three CDP programs for waveset distortion available on the CD-ROM: RESHAPE, OMIT and INTERACT. Whilst RESHAPE reads each wavecycle and replaces it with a different waveform of the same length, the OMIT program adulterates a sound file by omitting wavecycles, thus producing holes of silence in the sound. As for INTERACT, it alternatively interleaves the wavecycles of two different sounds. Use a Web browser to read the CDP user manual (file *Cdpman.htm*) for more information on these programs.

5.4 Statistical wavecycle synthesis

This technique works by synthesising waveforms from a sequence of interpolated breakpoint samples produced by a statistical formula; e.g. Gaussian, Cauchy or Poisson. The interpolation may be calculated using a number of functions, such as exponential, logarithmic and polynomial. Due to the statistical nature of the wavecycle production, this technique is also referred to as *dynamic stochastic synthesis*.

Figure 5.16 Gendy builds wavecycles by creating sequences of breakpoint values

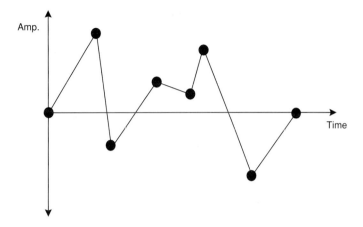

This technique has been developed by the composer Iannis Xenakis at Cemamu (Centre for Research in Mathematics and Music Automation) in France. Xenakis recently designed a synthesis system called Gendy (Serra, 1993). Gendy's primary assumption is that no wavecycle of a sound should be equal to any other, but rather that it should be in a constant process of transformation. The synthesis process starts with one cycle of a sound and then modifies its form each time the cycle is synthesised. The transformation is calculated by a probability distribution formula but the system does not apply the formula to every sample of the wavecycle. Instead, Gendy selects a much smaller amount of samples to forge a representation of the wavecycle as a sequence of breakpoints (Figure 5.16). At each cycle, the probability formula calculates new breakpoint values and the samples between are generated by an interpolation function. New coordinates for each breakpoint are constrained

Figure 5.17 In order to avoid complete disorder, new breakpoint values are generated using a probability function that narrows the result to a certain range of values

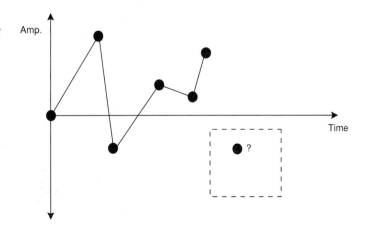

to remain within a certain boundary in order to avoid complete disorder (Figure 5.17). Also, different probability functions can be specified for each axis of the coordinate; i.e. one for the amplitude and one for the position of the breakpoint.

The parameters for *statistical wavecycle synthesis* are normally:

- Number of breakpoints in a wavecycle
- Type of probability function(s)
- Type of interpolation
- Boundary constraints for each breakpoint of the wavecycle

5.5 Sequential waveform composition

Sequential waveform composition works by letting the computer generate sample pattern sequences according to a user-defined formula. The patterns need not be a single wavecycle or a waveset. In fact, the whole sequence would not necessarily need to form any pattern at all. Some patterning is, however, desirable here because a minimum degree of repetition is indispensable for the production of a waveform. In other words, repetition is necessary because it determines the 'form' of a waveform, otherwise the outcome will always be white noise.

In essence, the technique resembles the statistical wavecycle synthesis technique introduced earlier, with the fundamental difference that here the 'breakpoints' are deterministic, in the sense that the segments are defined explicitly rather than statistically. Despite the deterministic nature of the technique itself, its outcome is often unpredictable, but notwithstanding interesting and unique. Like the binary instruction technique discussed in Chapter 3, synthesis by sequential waveform composition is also commonly referred to as *non-standard synthesis*. Indeed, in order to work with this technique one needs to be prepared to work with non-standard ways of thinking about sounds, because its synthesis parameters by no means bear a direct relation to acoustics.

The composer Arun Chandra, based in Illinois, USA, is one of the main supporters of the sequential waveform composition technique (Chandra, 1994). He has devised *Wigout* and *TrikTrak*, a unique pair of programs specifically designed for sequential waveform composition. Both programs, plus a number of examples and a comprehensive tutorial – specially prepared for this book by Arun Chandra himself – are available on the accompanying CD-ROM in folder *Wigtrik*.

In order to understand the fundamental principles of this technique one should look at the basics of digital sound synthesis

Figure 5.18 If the computer produces the same sample all the time, the result will be silence

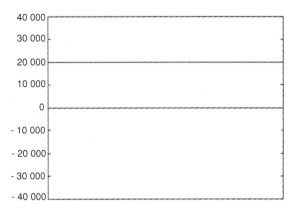

introduced in Chapter 1 from a slightly different angle. To reiterate, in order to produce a sound the computer feeds a series of numbers, or samples, to a digital-to-analog converter (DAC). The range of numbers that a DAC can manage depends upon the size of its word (refer to Chapter 1). For example, a 16-bit DAC handles 65 536 different values, ranging from –32 768 to +32 767 (zero is included). The number of sample values the DAC converts in one second is referred to as the sampling rate. For instance, a sampling rate equal to 44 100 Hz means that the DAC converts 44 100 samples (i.e. numbers between –32 768 and +32 767) per second.

In order to produce musically useful sounds on a computer, it must be programmed to produce the right sequence of samples. If the computer keeps producing the same sample value (e.g. 20 000), the result will be silence (Figure 5.18). Conversely, if the computer is programmed to produce sequences of different streams of sample values, then the output will be audible. It is exactly at this level that Wigout and TrikTraks work. For example, suppose that the computer is set to produce sequences of four different sample streams, as follows:

1   A stream of 20 samples equal to +30 000 followed by
2   A stream of 30 samples equal to –20 000 followed by
3   A stream of 10 samples equal to +16 000 followed by
4   A stream of 40 samples equal to –5000.

In this case, the computer produces the waveform depicted in Figure 5.19. The whole sequence (referred to as a *state*, in Wigout terms) is composed of 100 samples in total. Thus in order to produce a one-second sound at a sampling rate of 44 100 Hz, the computer must repeat this 'state' 441 times (441 × 100 = 44 100).

In Wigout and TrikTrak terms, each stream of homogeneous samples is referred to as an *element*. In the case of Figure 5.19,

Figure 5.19 A typical Wigout waveform

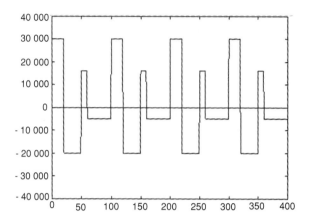

the waveform is specified in terms of four elements; each element is specified by two parameters: amount of samples (or 'duration') and amplitude (in this case a value between –32 768 and +32 767):

- Element 1: a stream of 20 samples equal to +30 000
- Element 2: a stream of 30 samples equal to –20 000
- Element 3: a stream of 10 samples equal to +16 000
- Element 4: a stream of 40 samples equal to –5000

Essentially, the technique works by specifying a collection of elements and by arranging these elements in a sequence that together comprises the waveform; hence the name: *sequential waveform composition*. A variation of the waveform portrayed in Figure 5.19 could be defined using the same four elements as follows (Figure 5.20): element 2 + element 4 + element 1 + element 3 + element 1 + element 4 + element 3.

Figure 5.20 A variation of the waveform depicted in Figure 5.19

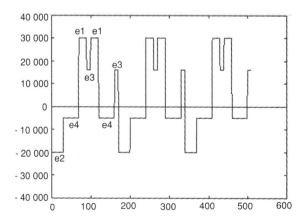

Figure 5.21 Besides squared shapes, Wigout also supports triangular shapes, curved shapes and a combination of all three

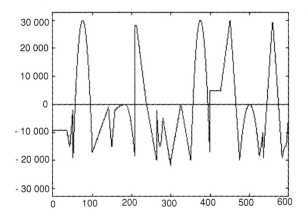

A great variety of waveforms can be produced using this technique. Wigout allows for the specification of waveforms that change the duration and the amplitude of their elements during production. Also, besides squared shapes one can also define triangular shapes, curved shapes and a combination of all three (Figure 5.21). TrikTraks complements the functionality of Wigout by adding the means for the specification of more complex waveform transitions; e.g. using non-linear functions.

6 Spectral modelling techniques

Until recently, the supremacy of pitch over other sound attributes prevailed in the theory of Western music. Time and timbre, for example, were considered to be of secondary importance. Today, however, this ideology is changing, and the tendency is now to consider that attributes such as pitch, time and timbre are all descriptors for the same phenomenon, but from different perspectives. For example, pitch may be considered as one aspect of the perception of timbre and timbre as one aspect of the perception of time. Pitch may, in turn, be considered as a quantifier to measure time. The music and theoretical work of composers such as Karlheinz Stockhausen, Iannis Xenakis, Pierre Schaeffer, Horacio Vaggione and Trevor Wishart have helped to emancipate the role of time and timbre in Western music (Schaeffer, 1966; Stockhausen, 1959; Vaggione, 1993; Wishart, 1985, 1994; Xenakis, 1971).

Western musicians have traditionally worked with conceptual models of individual instruments, which have specific boundaries of abstraction that give very little room for the significant manipulation of timbre. In the quest for methods to define new conceptual models that consider the importance of timbre in music composition, many composers and researchers have sought inspiration from psychological theories of sound perception. This includes the outcome of two relatively new scientific fields, namely psychoacoustics and cognitive sciences (Deutch, 1982; McAdams and Deliege, 1985). In this context a new paradigm for sound synthesis has naturally emerged: *spectral modelling*. Spectral modelling techniques employ parameters that tend to describe the sound spectrum, regardless of the acoustic mechanisms that may have produced them.

Spectral modelling techniques are the legacy of the *Fourier analysis* theory. Originally developed in the nineteenth century, Fourier analysis considers that a pitched sound is made up of various sinusoidal components, where the frequencies of higher components are integral multiples of the frequency of the lowest component. The pitch of a musical note is then assumed to be determined by the lowest component, normally referred to as the *fundamental frequency*. In this case, timbre is the result of the presence of specific components and their relative amplitudes, as if it were the result of a chord over a prominently loud fundamental with notes played at different volumes. Despite the fact that not all interesting musical sounds have a clear pitch and the pitch of a sound may not necessarily correspond to the lower component of its spectrum, Fourier analysis still constitutes one of the pillars of acoustics and music.

6.1 Additive synthesis

Additive synthesis is deeply rooted in the theory of Fourier analysis. The technique assumes that any periodic waveform can be modelled as a sum of sinusoids at various amplitude envelopes and time-varying frequencies. An additive synthesiser hence functions by summing up individually generated sinusoids in order to form a specific sound (Figure 6.1). Gottfried Michael Koenig was one of the first composers to create a piece entirely using additive synthesis: '*Klangfiguren I*', composed in 1955 in Cologne, Germany.

Additive synthesis is accepted as perhaps the most powerful and flexible spectral modelling method. However, this technique is difficult to control manually and is expensive to use on a computer. Musical timbres are composed of dozens of time-varying partials, including harmonic, non-harmonic and noise components. It would require dozens of oscillators, noise generators and envelopes to obtain convincing acoustic simulations using the additive technique. The specification and control of the parameter values for these components are also difficult and time consuming. Alternative methods have been proposed to improve this situation by providing tools to obtain automatically the synthesis parameters from the analysis of the spectrum of sampled sounds.

Figure 6.1 Additive
synthesis functions by
summing up sinusoids in
order to form specific
waveforms

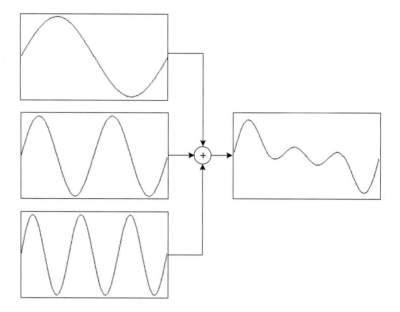

Figure 6.1 Additive synthesis functions by summing up sinusoids in order to form specific waveforms

Although additive synthesis specifically refers to the addition of sinusoids, the idea of adding simple sounds to form complex timbres dates back to the time when people started to build pipe organs. Each pipe produced relatively simple sounds that combined to form rich spectra. In a way, it is true to say that the organ is the precursor of the synthesiser.

On the accompanying CD-ROM, Virtual Waves (in folder *Virwaves*) provides a module called *Additive Synthesis* (Figure 6.2) that facilitates the creation of additive instruments (e.g. *Addbass1.syn* and *Addbass2.syn*); Kenny McAlpine's tutorial provides a step-by-step instruction to building additive synthesis instruments in Audio Architect (in folder *Audiarch*). For greater flexibility and control, Som-A (in folder *Soma*) is a programming language specifically designed for additive synthesis. In addition, Nicky Hind's CLM and Robert Thompson's pcmusic tutorials also provide some examples.

6.2 An introduction to spectrum analysis

Spectrum analysis is of fundamental importance for spectral modelling because samples alone do not inform the spectral constituents of a sampled sound. In order to model the spectrum of sounds, musicians need adequate means to dissect, interpret and represent them. A number of methods have been created to analyse the spectrum of sounds. There are two categories of spectrum analysis: *harmonic analysis* and *formant analysis*. Whilst

Figure 6.2 The *Additive Synthesis* module of Virtual Waves facilitates the creation of additive synthesisers

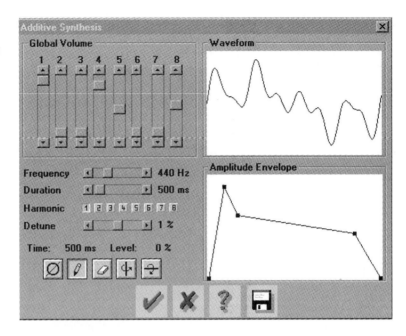

the former category is aimed at the identification of the frequencies and amplitudes of the spectrum components (Figure 6.3(a)), the latter uses the estimation of the overall shape of the spectrum's amplitude envelope (Figure 6.3(b)). *Short-time Fourier transform* (STFT) and *wavelet analysis* are typical examples of harmonic analysis, and *predictive analysis* is a typical example of formant analysis. Both categories have their merits and limitations; as far as sound synthesis is concerned, there is no optimum analysis technique. Some may perform better than others, according to the nature of the task at hand.

6.2.1 Short-time Fourier transform

Short-time Fourier transform (STFT) stands for an adaptation, suitable for computer programming, of the original Fourier analysis mathematics for calculating harmonic spectra. It is important to bear in mind that there are at least four other terms that are closely related to STFT, which are prone to cause confusion: *Fourier transform* (FT), *discrete-time Fourier transform* (DTFT), *discrete Fourier transform* (DFT) and *fast Fourier transform* (FFT). An in-depth discussion of the differences between these terms is beyond the scope of this book. On the whole, they differ in the way that they consider time and frequency, as follows:

1   FT is the original Fourier analysis mathematics whereby time and frequency are continuous

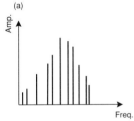

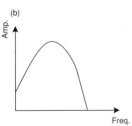

Figure 6.3 The two categories of spectrum analysis: (a) harmonic analysis and (b) formant analysis

2 DTFT is a variation of FT in which time is discrete, but frequency is continuous

3 DFT is a further version of FT in which both time and frequency are discrete

4 FFT is a faster version of DFT especially designed for computer programming

Fourier analysis detects the harmonic components of a sound using a pattern-matching method. In short, it functions by comparing a self-generated *virtual signal* with an input signal in order to determine which of the components of the former are also present in the latter. Imagine a mechanism whereby the components of the input signal are scanned by multiplying it by a reference signal. For instance, if both signals are two identical sinewaves of 110 Hz each, then the result of the multiplication will be a sinusoid of 220 Hz, but entirely offset to the positive domain (Figure 6.4). The offset value depends upon the amplitudes of both signals. Thus, it is also possible to estimate the amplitude of the input signal by taking the amplitude of the virtual signal as a reference.

The mathematics of the Fourier transform suggest that the harmonics of a composite signal can be identified by the occurrences of matching frequencies, whilst varying the frequency of the reference sinewave continuously. The digital version of this method may be simulated on a computer by scanning the input signal at rates that are multiples of its own fundamental frequency. This is basically how FFT works.

Figure 6.4 The result of the multiplication of two identical sinusoid signals will be a sinusoid entirely offset to the positive domain

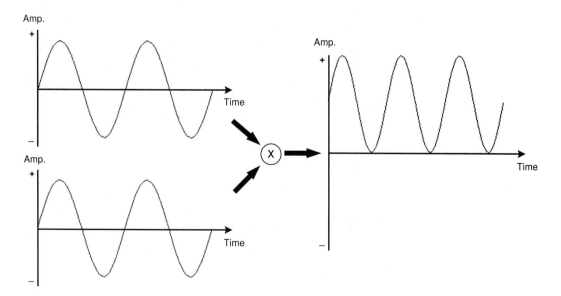

One of the main problems with the original Fourier transform theory is that it does not take into account that the components of a sound spectrum vary substantially during its course. In this case, the result of the analysis of a sound of 5 minutes duration, for example, would inform the various components of the spectrum but would not inform when and how they developed in time. For instance, the analysis of a sound that starts as a flute and changes its timbre to a violin during its course would only display the existence of the components that form both timbres, as if they were heard simultaneously. Short-time Fourier transform implements a solution for this problem. It chops the sound into short segments called windows and analyses each segment sequentially. It normally uses FFT to analyse these windows, and plots the analysis of the individual windows in sequence in order to trace the time evolution of the sound. The result of each window analysis is called an *FFT frame*. Each FFT frame contains two types of information: a *magnitude spectrum* depicting the amplitudes of every analysed component and a *phase spectrum* showing the initial phase for every frequency component.

In order to understand the functioning of the FFT mechanism, imagine a filtering process in which a bank of *harmonic detectors* (e.g. band-pass filters) are tuned to a number of frequencies (see Chapter 4 for more about filters). The greater the number of harmonic detectors, the more precise the analysis. For instance, if each detector has a bandwidth of 30 Hz, it would be necessary to employ no less than 500 units to cover a band ranging from 30 Hz to 15 000 Hz. Such a bandwidth is appropriate for lower-frequency bands where the value of a semitone is close to 30 Hz. However, this bandwidth value is unnecessarily narrow at higher-frequency bands, where 30 Hz can be as small as one tenth of a semitone. Unfortunately, FFT distributes its harmonic detectors linearly across the audio range. This means that a compromise has to be found in order to prevent unnecessary precision at higher-frequency bands, on the one hand, and poor performance at lower-frequency bands, on the other.

Windowing

In the context of STFT, windowing is the process whereby shorter portions of a sampled sound are detached for FFT analysis (Figure 6.5). The windowing process must be sequential, but the windows may overlap (Figure 6.6). The effectiveness of the STFT process depends upon the specification of three windowing factors: the *envelope for the window*, the *size of the window* and the *overlapping factor*.

Figure 6.5 Windowing is a process in STFT whereby shorter portions of the sound are detached for FFT analysis

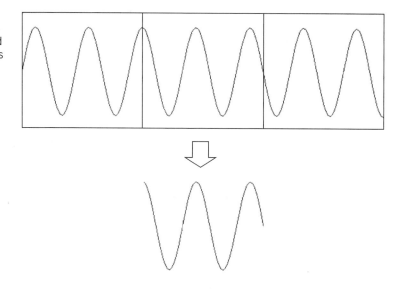

Figure 6.6 Overlapping windows

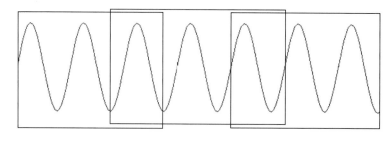

Figure 6.7 Cutting the sound at non–zero parts causes irregularities in the analysis

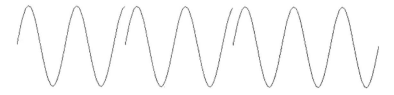

Note in Figure 6.5 that the windowing process may cut the sound at non-zero parts of the waveform. The FFT algorithm considers each window as a unit similar to a wavecycle. The problem with this is that interruptions between the ends of the windowed portion lead to irregularities in the analysis (Figure 6.7). This problem can be remedied by using a lobe envelope to smooth both sides of the window (Figure 6.8). From the various functions that generate lobe envelopes, the *Gaussian*, the *Hamming* and the *Kaiser* functions are more often used because they tend to produce good results.

Figure 6.8 Interruptions between the end parts of the windowed portion can be remedied by using a lobe envelope to smooth both sides of the window

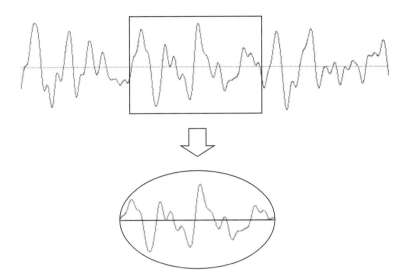

Lobe envelopes, however, have their own malign side-effects. Note that the amplitude of the whole signal now rises and falls periodically. This is a clear case of *ring modulation* (see Chapter 3) in which the 'windowing frequency' acts as the modulating frequency. Enveloped windowing also may distort the analysis process, but the result is still more accurate than with windowing without envelopes.

The size of the window defines the frequency resolution and the time resolution of the analysis. This value is normally specified as a power of two; e.g. 256, 512, 1024, etc. Longer windows have better frequency resolution than smaller ones, but the latter have better time resolution than the former. For example, whilst a window of 1024 samples at a rate of 44 100 Hz, allowing for a time resolution of approximately 23 milliseconds (1024/44 100 = 0.023), a window of 256 samples allows for a much better resolution of approximately 6 milliseconds (256/44 100 = 0.0058). Conversely, the harmonic detectors will be tuned to scan frequencies spaced by a bandwidth of approximately 43 Hz (44 100/1024 = 43) in the former case and to approximately 172 Hz (44 100/256 = 172) in the latter. This means that a window of 256 samples is not suitable for the analysis of sounds lower than 172 Hz (approximately an F3 note), but it may suit the analysis of a sound that is likely to present important fluctuation within less than 23 milliseconds.

To summarise, in order to know precisely when an event occurs, the FFT algorithm cuts down the frequency resolution and vice versa. The overlapping of successive windows is normally employed to alleviate this controversy. For example, if an overlap factor is set to equal 16 (i.e. 1/16th of the size of the window) and the window size is set to equal 1024 samples, then the windowing process will slice the sound in steps of 64 samples (i.e. 1024/16 = 64). In this case the time resolution of a window of 1024 samples will improve from 23 milliseconds to approximately 1.4 milliseconds (i.e. 0.023/16 = 0.0014).

6.2.2 Wavelets analysis

The size of the STFT window is constant. Hence all harmonic detectors have the same bandwidth and are placed linearly across the audio range. The wavelets method improves this situation by introducing a mechanism whereby the size of the window varies according to the frequency being analysed. That is, the bandwidths of the harmonic detectors vary with frequency and they are placed logarithmically across the audio range.

The wavelets method is inspired by a concept known as constant Q filtering. In filters jargon, Q denotes the ratio between the band-pass filter's resonance centre frequency and its bandwidth. If Q is kept constant, then the bandwidths of the BPF bank vary according to their resonance centre frequency.

As far as wavelets analysis is concerned, the centre frequency of a BPF compares to the size of the window. Thus the length of the analysis window varies proportionally to the frequencies being analysed. This mechanism effectively minimises the resolution controversy of the STFT analysis because it intensifies both time resolution at high-frequency bands and frequency resolution at low-frequency bands.

Wavelets, however, demand powerful computation. Operative implementations of this analysis method are only currently available on powerful computers. Since personal computers are becoming increasingly fast, the next edition of this book is bound to present a number of new efficient analysis and resynthesis tools based upon wavelets.

6.2.3 Predictive analysis

Harmonic analysis methods, such as STFT and wavelets, are not entirely adequate for the analysis of sounds with a high proportion of non-sinusoidal components and, to a certain extent, to non-harmonic combinations of partials. The nature of these signals is not compatible with the notion that sounds are composed of harmonically related and stable sinusoids. Formant analysis proposes an alternative method of representation which is very similar to the functional approach to physical modelling, discussed in Chapter 4. The sound is represented here in terms of an overall predictive mould that shapes a signal rich in partials, such as a pulse wave or white noise. The advantage of this method is that the predictive mould does not need to specify the frequencies of the spectrum precisely; any value within a certain range may qualify.

Predictive analysis is a typical example of formant analysis. The core of this method is the interplay between two kinds of filters: the *all-pole filter* and the *all-zero filter*. In electronic engineering jargon, the pole of a filter refers to a point of resonance in the spectrum and the zero of a filter to a point of attenuation. An all-pole filter is a filter that allows for several resonant peaks in the spectrum. Conversely, an all-zero filter creates various notches in the spectrum.

The main target of predictive analysis is to create an all-zero filter that corresponds to the inverse of the acoustic mould that originally shaped the spectrum of the sound in question. The analysis algorithm first reads snapshots of samples and estimates time-varying coefficients for an all-pole filter that could recreate them. These estimated coefficients are then inverted to fit an all-zero filter. Next, the all-zero filter is applied to the signal in order to test the accuracy of the estimated parameters. In theory, the all-zero filter should cancel the effect of the acoustic mould; i.e. the result of the all-zero filtering should be equivalent to the stimulation signal. The accuracy of the all-zero filter is measured by comparing its output with the original stimulation. This process is repeated until all samples of the sound have been analysed.

The main criticism of predictive analysis is that the original stimulation signal is rarely known. In practice, the algorithm predicts (*sic*) that the stimulation is either a pitched pulse train or white noise. In other words, the outcome of the all-zero filter is compared to an artificial stimulation that is rather rudimentary, thus limiting the efficiency of technique to specific types of sounds; e.g. prediction analysis works satisfactorily for speech and some wind sounds.

6.3 Analysis and resynthesis

Most spectral modelling techniques work in two stages – analysis and resynthesis – which can be performed in a number of

Figure 6.9 Spectral time stretching works by enlarging the distances between the frames of the original sound (a) in order to stretch the spectrum of the sound (b)

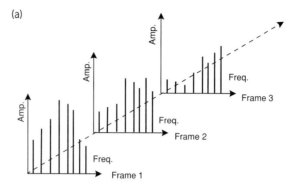

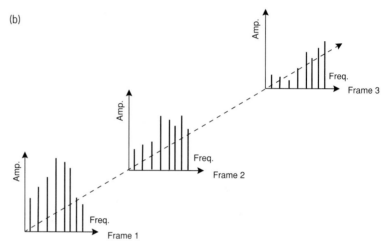

Figure 6.10 Spectral shift

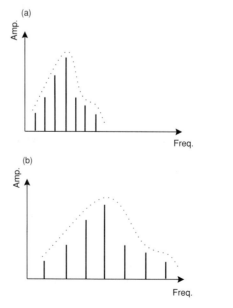

Figure 6.11 Spectral stretching

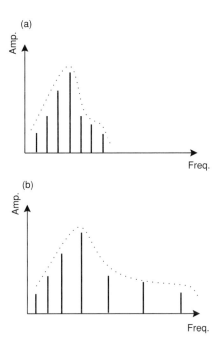

ways. As has already been demonstrated, the analysis stage extracts a number of parameters from a sampled sound. The resynthesis stage then uses these parameters to recreate the sound using a suitable synthesis technique. The ability to create a precise imitation of the original sound is not, however, very interesting for composers who obviously would prefer to modify the analysis data in order to synthesise variants of the original sound. The great advantage of the analysis and resynthesis technique over plain sampling is that analysis data can be modified in a variety of ways in order to create new sounds. Typical modifications include spectral time stretching (Figure 6.9), spectral shift (Figure 6.10) and spectral stretching (Figure 11).

6.3.1 Resynthesis by Fourier analysis reversion

The resynthesis by Fourier analysis reversion method (sometimes referred to as *overlap-add* resynthesis) takes the information from each frame of the STFT analysis and recreates an approximation of the original sound window by window. This process works well for those cases where the windows are reconstructed in conformity with the original analysis specification (e.g. use of similar windowing and overlapping). The analysis data must be cautiously manipulated in order to produce good results, otherwise the integrity of the signal may be completely destroyed. This may, of course, be desirable for some specific purposes, but in most circumstances such spurious effects are difficult to control. This resynthesis method is of limited interest from the standpoint of a composer who may require greater flexibility to transform the sound.

6.3.2 Additive resynthesis

This method employs STFT analysis data to control an additive synthesiser. In this case, the analysis stage includes an algorithm that converts the STFT data into amplitude and frequency trajectories – or envelopes – that span across the STFT frames. This is generally referred to as the *peak-tracking algorithm* and can be implemented in a number of ways.

The additive resynthesis method is more flexible for parametric transformation than the analysis reversion technique discussed above, because envelope data are generally straightforward to manipulate. Transformations such as stretching, shrinking, rescaling and shifting can be successfully applied to either or both frequency and time envelopes. This resynthesis method is suitable for dealing with harmonic sounds with smoothly changing, or almost static, components, such as a note played on a clarinet. Noisy or non-harmonic sounds with highly dynamic spectra, such as rasping voices, are more difficult to handle here, but not impossible. This limitation of additive resynthesis is due to the fact that STFT analysis data is somehow adapted (or 'quantised' in music technology jargon) to fit the requirements for conversion into envelopes. Some information needs be discarded in order to render the method practical, otherwise it would be necessary to use a prohibitively vast additive synthesiser to reproduce the original signal.

The Phase Vocoder system implemented by the composer Trevor Wishart and his collaborators at Composer's Desktop Project (CDP) in England is a classic example of additive resynthesis. In order to operate the Phase Vocoder the musician first needs to define a number of parameters, including the *envelope for the window*, the *size of the window* and the *overlap factor*.

From the various types of envelopes for windowing available, Phase Vocoder normally employs either the *Hamming function* or the *Kaiser function*. Both functions have a bell-like shape. One may provide better analysis results than the other, depending upon the nature of the sound analysed and the kind of result expected.

The size of the window defines the number of input samples to be analysed at a time. The larger the window, the greater the

Figure 6.12 The Phase Vocoder uses channels to act as 'harmonic detectors'. The greater the number of channels, the better the frequency resolution. Note that (b) does not distinguish between the two uppermost partials that are perfectly distinct in (a)

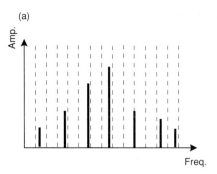

(a)

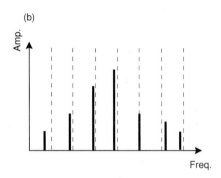

(b)

number of channels, but the lower the time resolution, and vice versa. This should be set large enough to capture four periods of the lowest frequency of interest. The sampling rate divided by the number of channels should be less than the lowest pitch in the input sound. This may be set in terms of the *lowest frequency* of interest or in terms of the *number of channels required*. Note that the term 'channel' is the Phase Vocoder jargon for what we referred earlier to as harmonic detectors (Figure 6.12).

The overlap factor is the number of samples that the algorithm skips along the input samples each time it takes a new window (Figure 6.13). This factor is specified as a fraction of the window size; e.g. an eighth of the window size.

The real power of the CDP's Phase Vocoder package is the great number of tools available for spectral manipulation techniques (Wishart, 1994). There are currently over 60 tools in the package for tasks such as:

- Blurring the spectrum
- Amplifying or attenuating the whole spectrum
- Interpolating (morphing) different sounds
- Shuffling analysis windows
- Stretching the spectrum

137

Figure 6.13 In this case a window has 512 samples and the overlap factor is set to 64 samples (512/8 = 64)

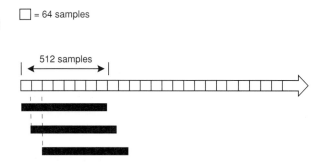

□ = 64 samples

512 samples

- Imposing the spectral envelope from one sound onto another

On the accompanying CD-ROM, in folder *Cdp*, there are seven programs from the CDP Phase Vocoder suite to perform the tasks listed above: CDPVOC, BLUR, GAIN, MORPH, SHUFFLE, STRETCH and VOCODE.

CDPVOC is the actual program that performs the task of analysis and resynthesis; the remaining six are programs for spectral manipulations. Note that the spectral manipulation programs do not process a sound directly but the information generated by the analysis process. In order to manipulate the spectrum of the sound one must generate its analysis file first. The analysis file can then be submitted to any of the spectral transformation programs. The transformation process will always produce another analysis file, which in turn must be resynthesised in order to be heard; that is CDPVOC must return to the scene in order to resynthesise the sound.

The BLUR program blurs the spectrum of the sound by time-varying its components and the GAIN program amplifies or attenuates the whole spectrum envelope according to a user-specified factor. MORPH is a program which creates transitions between two sounds. SHUFFLE, as its name suggests, shuffles the order of the analysis windows. The STRETCH program is used to stretch the frequencies in the spectrum; that is, it expands or contracts the distances between the partials (Figure 6.11). Finally, VOCODE reshapes a sound with the characteristics of another sound, by imposing the spectral contour of the latter onto the former. Refer to the documentation on the CD-ROM for more details of the operation of these programs and for some examples.

Also on the CD-ROM there is a demonstration version of Diphone for Macintosh. Diphone is a rather sophisticated program for making transitions between sounds (see Chapter 2).

In addition, Raffaele de Tintis' Morph for PC (in folder *Morph*) offers an alternative tool for making sound hybrids.

6.3.3 Subtractive resynthesis

Subtractive resynthesis takes a rather different route from additive resynthesis. It employs formant analysis to produce the filter coefficients for a subtractive synthesiser. The advantage of the subtractive paradigm over the additive one is that it considers pitch and spectrum independently of each other. That is, in subtractive synthesis the shape of the spectrum is not subjected to a fundamental frequency, in the sense that the pitch of the stimulation signal does not affect the resonator. Note, however, that there is no guarantee that the stimulation signal will contain the components required to resonate.

Linear predictive coding (LPC) is a typical example of subtractive resynthesis employing predictive analysis. The result of the predictive analysis stage is a series of snapshots, called *LPC frames*, which contain the information required to resynthesise the sound. This information may vary in different implementations of the LPC algorithm, but it generally includes the following:

Figure 6.14 In LPC synthesis, analysis frames are read sequentially and the algorithm has to decide whether the stimulation should be a noise signal or a pulse stream

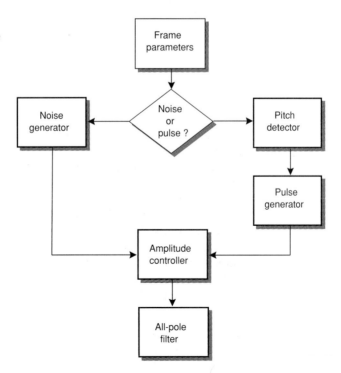

- The length of the frame (in samples or seconds)
- The average amplitude of the frame's original signal
- The average amplitude of the estimated output from the inverted filter
- The pitch for the frame
- The coefficients for the all-pole filter

The synthesis stage is depicted in Figure 6.14. Frames are read sequentially and the algorithm first decides whether the stimulation should be a noise signal or a pulse stream. This decision depends upon the ratio between the amplitude of the original signal and the amplitude of the estimated output from the inverted filter. A large ratio (e.g. higher that 0.25) leads to a choice for noise stimulation. The stimulation is then scaled to the amplitude required and fed into the all-pole filter, which may have as many as sixty poles.

The LPC analysis frames may, of course, be edited in order to transform the sound. For instance, the duration of the frames may be stretched or shrunk in order to alter the duration of the sound without changing its pitch.

6.3.4 Resynthesis by reintegration of discarded components

This method attempts to improve additive resynthesis by adding a mechanism that reintegrates the information discarded during the STFT analysis. Each frame of the STFT is resynthesised immediately after the analysis, using the additive resynthesis method described above (Figure 6.15). This signal is then subtracted from the original in order to obtain the *residual* part that has been discarded or converted into stationary vibrations during the analysis process. This residual part is then represented as an envelope for a time-varying filter.

The outcome of the analysis process provides two branches of information: about the *sinusoidal components* of the sound and about *noisy and other non-sinusoidal components*. Note that this resynthesis method takes advantage of both categories of analysis discussed earlier: harmonic analysis and formant analysis. The type of representation generated by a formant analysis technique is highly suitable for representing the stochastic nature of the residual part.

The resynthesis process results from two simultaneous synthesis processes: one for sinusoidal components and the other for the noisy components of the sound (Figure 6.16). The sinusoidal components are produced by generating sinewaves dictated by

Figure 6.15 The algorithm deduces the residual part of the sound by subtracting its sinusoidal components

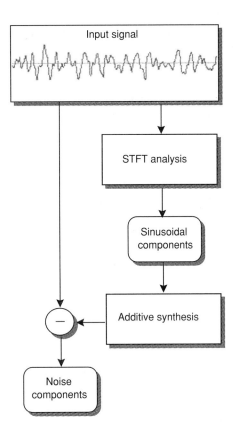

Figure 6.16 The resynthesis by reintegration of discarded components works by employing two simultaneous synthesis processes: one for the sinusoidal components and the other for the noisy part

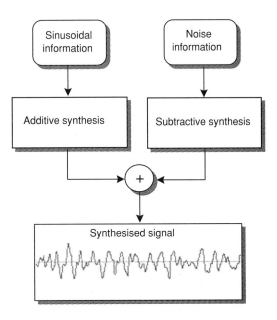

the amplitude and frequency trajectories of the harmonic analysis, as with *additive resynthesis*. Similarly, the 'stochastic' components are produced by filtering a white noise signal, according to the envelope produced by the formant analysis, as with *subtractive synthesis*. Some implementations, such as the SMS system discussed below, generate 'artificial' magnitude and phase information in order to use the Fourier analysis reversion technique to resynthesise the stochastic part.

The *Spectrum Modelling Synthesis* system (SMS) developed by Xavier Serra, of Pompeu Fabra University, in Barcelona, Spain, is an edifying example of resynthesis by reintegration of discarded components (Serra and Smith, 1990). The system package provides a number of modules and tools, including an analysis module and a synthesis module. The latter embeds facilities to modify the analysis data for resynthesis.

The analysis module has been designed to cope with a variety of different sounds, but it needs to be customised to the specific sound in question. In order to obtain a good analysis the musician must specify a number of parameters very carefully, such as:

- The format of the analysis; i.e. the algorithm is 'tuned' to target either a harmonic or a non-harmonic sound
- The STFT parameters such as window size and overlap factor
- The peak detection and peak continuation parameters such as the lowest and highest frequencies to be detected, and the spectral magnitude threshold
- Stochastic analysis parameters such as the maximum number of breakpoints for the formant envelope

The synthesis module provides a number of controllers to transform the analysis data for resynthesis. Here the musician can specify a large range of spectral transformations, such as:

- To scale the amplitude of the overall sound, specific areas of the spectrum or individual components
- To scale the frequencies of the overall spectrum, specific areas of the spectrum or individual components
- To modify the distribution of the harmonics by stretching or compressing them
- To amplitude modulate and frequency modulate the components of the spectrum
- To change the envelope of the residual components

Powerful and flexible synthesis systems are not always easy to master and the SMS system is no exception to this rule. It is fundamental to work with good analysis data in order to produce interesting sounds here. The two main requisites to effectively operate the SMS program are patience and experience.

On the CD-ROM, in folder *Sms*, there is a version of Xavier Serra's SMS for PC-compatible under Windows 95 enriched with examples and comprehensive documentation.

6.4 Formant synthesis

Formant synthesis originated from research into human voice simulation, but its methods have been adapted to model other types of sounds. Human voice simulation is of great interest to the industrial, scientific and artistic communities alike. In telecommunications, for example, speech synthesis is important because it is comparatively inexpensive and also safer to transmit synthesis parameters rather than the voice signal. In fact, music sound synthesis owes much of its techniques to research in telecommunications systems.

In music research, voice simulation is important because the human ability to distinguish between different timbres is closely related to our capacity to recognise vocal sounds, especially vowels. Since the nineteenth century the bulk of scientific studies have linked speech and timbre perception, from Helmholtz (1885) to current research in cognitive science (McAdams and Deliege, 1985).

The desired spectrum envelope of human singing has the appearance of a pattern of 'hills and valleys', technically called formants (Figure 6.17). The mouth, nose and throat function as

Figure 6.17 The desired spectrum envelope of human singing has the appearance of a pattern of 'hills and valleys', technically called formants

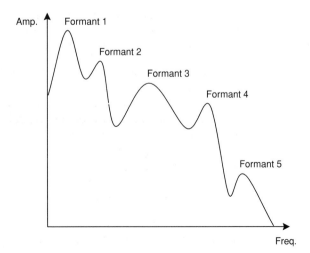

a resonating tube whereby particular frequency bands are emphasised and others are attenuated. This resonating system can be simulated by a composition of parallel band-pass filters (BPF), in which each band-pass centre frequency is tuned to produce a formant peak. In the context of spectral modelling, however, formant synthesis does not attempt to build such a model. The core purpose of formant synthesis is to produce the formants using dedicated generators in which musicians specify the spectrum of the required sound, rather than the mechanism that produces it. One of the main advantages of this approach is that generators specifically designed to produce formants require less computation than their filter counterparts.

6.4.1 FOF generators

One of the most successful formant generators currently available is the FOF generator devised in the early 1980s by Xavier Rodet at Ircam, in Paris. In fact, 'FOF' is an acronym for 'Function d'Onde Formantique' (Rodet, 1984). Briefly, a single FOF unit replaces the subtractive synthesis pair *stimulation function* followed by a *band-pass filter*. The signal produced by the FOF generator is equivalent to the signal generated by a pulse-driven filter.

The generator works by producing a sequence of dampened sinewave bursts. A single note contains a number of these bursts. Each burst has its own local envelope with either a steep or a smooth attack, and an exponential decay curve. The formant is the result of this local envelope. As the duration of each FOF burst lasts for only a few milliseconds, the envelope produces sidebands around the sinewave, as in *amplitude modulation*. Note that this mechanism resembles the granular synthesis technique, with the difference that the envelope of the 'FOF grain' was especially designed to facilitate the production of formants.

The FOF generator is controlled by a number of parameters, including amplitude, frequency and local envelope. The only difficulty with the FOF generator is to relate the shape of the envelope with the required formant (Figure 6.18). There are four formant parameters and two of them refer to the attack and decay of the local envelope:

- Formant centre frequency
- Formant amplitude
- Rise time of the local envelope
- Decay time of the local envelope

Figure 6.18 The shape of the FOF envelope determines the formant produced by the generator

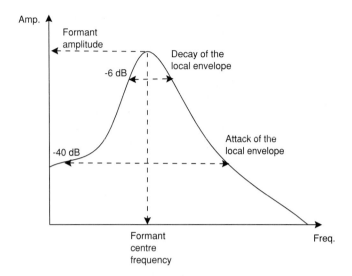

The decay time local envelope defines the bandwidth of the formant at –6 dB and the rise time the skirtwidth of the formant at –40 dB. The relationships are as follows: the longer the decay, the sharper the resonance peak, and the longer the rise time, the narrower the skirtwidth. Reference values for singing sounds are approximately 7 milliseconds for the rise time and 3 milliseconds for the decay time.

This technique was originally designed for a synthesis system developed by Ircam in Paris, called *Chant* (in French 'chant' means sing). Some languages and systems provide a FOF unit generator as part of their pool of modules; e.g. Virtual Waves introduced in Chapter 2.

6.4.2 Vosim synthesis

Vosim is an abbreviation of voice simulator. This synthesis technique was developed in the 1970s by Werner Kaegi and Stan Tempellars (1978) at the Institute of Sonology, in Utrecht, Holland. In plain terms, this technique is akin to FOF, in the sense that it produces a formant by generating sequences of tone bursts. The main difference lies in the waveform of the bursts.

A Vosim unit generates a series of bursts containing a number of pulses decreasing in amplitude. Each pulse is the result of the square of a sinewave (i.e. $\sin^2$). The *width* of each pulse determines the *centre frequency of the formant* and the end of each burst is marked by a gap which determines the *frequency of the burst*.

As with FOF, it is necessary to use several Vosim units in order to create a spectrum with various formants. Each unit is controlled by setting a number of parameters including:

- Width of the pulse
- Gap size between the bursts
- Number of pulses per burst
- Amplitude of the first pulse
- Attenuation factor for the series of pulses

7 Instrument and sound design

One of the great strengths of computer sound synthesis is its potential for the creation of an infinite variety of instruments and sounds. So far this book has introduced a number of synthesis techniques to explore this potential. Ideally, the electronic musician should master all these techniques, but this is neither practical nor absolutely necessary for the production of good music. In practice, composers have an idea of what is generally possible and then select a few techniques with which to build their own modus operandi.

When designing synthesis instruments, we should give preference to those techniques that best fit the way we understand the sound phenomena we want to work with. For example, whilst those composers interested in timbral manipulation will be better off working with analysis and resynthesis techniques (Chapter 6), those working with scientific metaphors (e.g. dynamic systems, cellular automata, etc.) would certainly prefer to use granular synthesis methods (Chapter 5).

It is impossible to develop a set of definite rules for sound and instrument design, but we can, however, establish some guidelines. These should normally emerge from the study of the work of experienced composers, and, of course, from your own experience.

This chapter introduces an approach to instrument and sound design. This should by no means be considered as a set of rules to be followed strictly, but rather as a reference point for the outline of your own approach. The chapter begins with a discussion about the relationship between the sound phenomena and the techniques for synthesis implementation. Next, we study in greater detail the design of an example synthesis instrument and of some particular sounds.

7.1 Associating cause and effect

Unless you are prepared to adopt a highly irrational approach to sound design, you should always give preference to basing your new instruments on well-documented acoustic theories; see Howard and Angus (1996) for a good introduction to acoustics. Acoustic musical instruments are a reliable point of departure because the human ear is familiar with them. However, be wary here, because the expression 'point of departure' does not necessarily mean that you should imitate these instruments; rather, computer sound synthesis should aim at the production of new instruments.

The basic spectral formation of the sounds produced by acoustic musical instruments may be classified into three general categories: sounds with *definite pitch*, sounds with *indefinite pitch* and *noise*. The analysis of the spectrum of these sound reveals that:

- Sounds with definite pitches often have harmonic partials
- Sounds with indefinite pitches often have inharmonic partials
- Noises tend to contain a large number of random components covering the whole audible band

This is indeed a very simplistic way of categorising sounds. Nevertheless, it can be a good starting point if you can fit the sounds you want to synthesise into one of these three categories. This will help you to establish which techniques will be more appropriate to produce them. For example:

- Sawtooth and narrow pulses are good starting points to model the harmonic partials produced by vibrating strings and open air columns; e.g. strings, brass
- Square and triangular waves are good starting points to model sounds with only odd harmonic partials produced by closed air columns; e.g. woodwinds, organ
- Amplitude and frequency modulations are good starting points to model the non-hamonic partials of sounds produced by hanging bars, bells and drums

- Noise generators are the best starting point to model the highly random partial configurations of the sounds of winds, snare drums and air blasts through a cylinder

Also, the changes of the 'shape' of a sound event over its duration are vital clues for the identification of the acoustic mechanisms that could have produced the sound. Acoustic instruments are played by an excitation which sets it into vibration. The type of excitation defines the shape of the sound; that is, whether the sound is plucked, bowed or blown, for example. There are two general ways of producing vibrations on an acoustic instrument: by *continuous excitation* or by *impulsive excitation*.

Continuous excitation is when the instrument needs the constant application of force to produce a note. In this case the sound will be heard as long as the force is being applied (e.g. a clarinet). Sounds from continuous excitation-type instruments often have a relatively slow attack time and will sustain the loudness until the excitation stops. Most sounds produced by continuous excitation begin with a very simple waveform at its attack time. Then, an increasing number of partials appear and build up the spectrum until the sound reaches the sustain time. When the excitation ceases, the partials often fade out in the opposite order to that in which they have appeared.

Figure 7.1 The two-axis diagram inspired by Pierre Schaeffer's concept of sound maintenance is an excellent example of a framework for sound categorisation

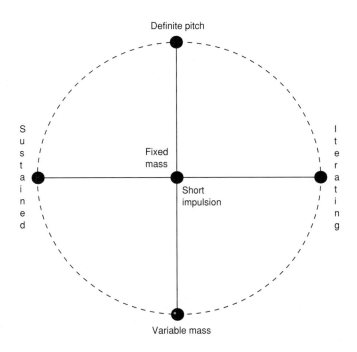

Impulsive excitation is when the instrument requires only an impulse to produce a note. In this case, the attack time is very short, almost instantaneous. The sound begins at its loudest and fullest instant, but it has virtually no sustain. As the sound fades its spectrum normally fades away in a descending order.

In *Traité des objets musicaux* Pierre Schaeffer (1966) purports to define criteria for the identification and classification of sounds according to their general features. Although Schaeffer was not explicitly concerned with sound synthesis itself, his work provides an excellent insight into making sound categorisations. One of the most important aspects of his theoretical work is the concept of *maintenance*, which involves the relationship between both criteria discussed above: the *basic spectral formation* and *mode of excitation*.

Schaeffer observed that sounds result from a certain energetic process, called *maintenance*, which describes the development which a sound undergoes with time. If the sound is merely ephemeral, a non-resonant single sound such as a drum stroke or a vocal plosive consonant, then there is a discrete short impulsion. If the sound is prolonged, such as a sung vowel, then there is a continuous, that is, sustained, sound. If it is prolonged by the repetition of impulsion, such as a drum roll, then there is iterating maintenance.

For a more systematic panorama, a two-axis diagram (Figure 7.1) can be defined. The middle horizontal axis contains sounds of *short impulsion*, the left features the *sustained sounds* and the right has those whose maintenance is *iterating*. On the vertical axis, sounds with *fixed mass* are placed between sounds of *definite pitches* and sounds of *variable mass*. Most sounds produced on acoustic musical instruments settle extremely fast into a fixed pitch or fixed mass (e.g. pink noise), following the attack stage. *Mass*, in this case, can be considered as the amount of information carried by the sound, ranging from redundancy (a meagre steady sinewave of fixed pitch) to total unpredictability (white noise). These perceptual phenomena may be implemented in a number of ways; for example by inputting noise into a band-pass filter whose passband is shortened (this tends to produce a sound with definite pitch) and widened (which results in a sound with a variable noise band).

The two-axis diagram in Figure 7.1 is an example of a starting point for the design of synthesis instruments. It provides an excellent framework for sound categorisation and encourages the specification of meaningful synthesis parameters; e.g. the bandwidth of the BPF mentioned above controls the mass of the sound.

Figure 7.2 In subtractive synthesis each formant is associated with the response of a band pass filter (BPF)

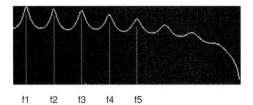

f1   f2   f3   f4   f5

7.2 Example study: synthesising human vocal sounds using subtractive synthesis

This section studies the design of an instrument that produces sounds resembling the human voice. The spectral contour of vocal-like sounds has the appearance of a pattern of 'hills and valleys' technically called formants (Figure 7.2). Among the various synthesis techniques capable of synthesising formants, the subtractive technique is used for this example. In subtractive synthesis each formant is associated with the response of a BPF; in this case, one needs a parallel composition of BPFs set to different responses. The signal to be filtered is simultaneously applied to all filters and the frequency responses of the filters are added together.

The example study instrument embodies the functional approach to physical modelling techniques discussed in Chapter 4. This approach considers that the behaviour of an instrument (in this case, the *human vocal tract*) is determined by two main components: *source* and *resonator*, where the former produces a raw signal that is shaped by the latter (Figure 7.3).

Figure 7.3 The example study instrument embodies the functional approach to physical modelling techniques

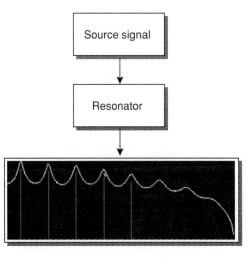

7.2.1 The source component

When singing or speaking, an airstream is forced upwards through the trachea from the lungs. At its upper end, the trachea enters the larynx, which in turn opens into the pharynx. At the base of the larynx, the vocal folds are folded inwards from each side, leaving a variable tension and a slit-like separation, both controlled by muscles in the larynx. In normal breathing, the folds are held apart to permit the free flow of air. In singing or speaking, the folds are brought close together and tensed. The forcing of the airstream through the vocal folds in this state sets them into vibration. As a result, the airflow is modulated at the vibration frequency of the vocal folds. Despite the fact that the motion of the vocal folds is not a simple but a non-uniform vibration, the pitch of the sound produced is determined by this motion. In order to simulate this phenomenon, our example instrument generates two types of sound sources: the *voicing source*, which produces a quasi-periodic vibration, and the *noise source*, which produces turbulence. The former generates a pulse stream intended to simulate the non-uniform (or quasi-periodic) vibration of the vocal folds, whereas the latter is intended to simulate an airflow past a constriction or a relatively wide separation of the vocal folds.

The voicing source

Figure 7.4 portrays the voicing source mechanism of our instrument. *Jitter* and *vibrato* are very important for voicing sound quality because they add a degree of non-uniformity to the fundamental frequency of the sound source. Jitter is defined as the difference in fundamental frequency from one period of the sound to the next. It normally varies at random between –6 per cent and +6 per cent of the fundamental frequency. Our instrument calculates this value by adding the results from three random number generators whose values are produced by interpolating periodic random values. The pcmusic code for the jitter generator is as follows:

```
ran b10 0.02 0.05Hz d d d;
ran b11 0.02 0.111Hz d d d;
ran b12 0.02 1.219Hz d d d;
adn b13 b10 b11 b12;
mult b14 b13 F0;              // b14=jitter factor
```

Vibrato is defined as the frequency modulation of the fundamental frequency. In general, vibrato is interesting from a

Figure 7.4 The voicing
source mechanism

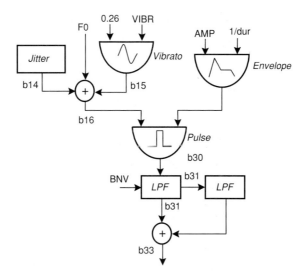

timbral point of view. For instance, in humans it is important for
the recognition of the identity of the singer. Typically, there are
two distinct parameters for vibrato control: the *width* of the
vibrato, that is, the amplitude of the modulating frequency, and
the *rate* of the vibrato (VIBR), the frequency of the modulating
frequency. VIBR is normally set to a value between 5.2 Hz and
5.6 Hz. If this value is set above the lower limit of the audio
range (approximately 18 Hz), then the effect will no longer be a
vibrato but a distortion (i.e. FM) – which may be useful to
produce some interesting effects. Vibrato is implemented using
an oscillator and is used in conjuction with jitter to produce
variations in the fundamental frequency, represented here as F0:

```
osc b15 0.26 VIBR f2 d;        // b15=vibrato
adn b16 b15 b14 F0;            // b16=freq+jitter+vibrato
```

The heart of the voicing source is a pulse generator. It produces
a periodic waveform at a specific frequency with a great deal of
energy in the harmonics. A pulse waveform has significant
amplitude only during relatively brief time intervals, called
pulse width (Figure 7.5). When a pulse waveform is periodically
repeated, the resulting signal has a rich spectrum. The output of
the voicing source provides the raw material from which the
filtering system will shape the required sound.

The human singer manages the source signal by adjusting the
muscles of the larynx, controlling the spacing and tension of the
vocal folds. In loud singing the vocal folds close completely for
a significant fraction of the vibration cycle. The resulting airflow

Figure 7.5 The spectrum of a pulse waveform has both odd and even harmonics

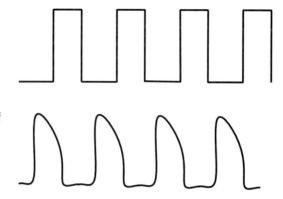

Figure 7.6 The simulation of a voicing source signal may be obtained by low-pass filtering the pulse train

is normally referred to as the *normal voicing* and it has a waveform like the one portrayed in Figure 7.6. If the folds are too slack, they do not close properly at any point in the vibration cycle. Here the resulting waveform loses its spiky appearance and starts to resemble a sinusoidal curve; this signal is referred to as the *quasi-sinusoidal* source. This results in the breathy sound quality often found among untrained singers. The expression 'trained singer' here denotes one with the ability to maintain the vocal folds pressed tightly together.

In order to simulate the production of both types of voicing signals, the pulse train is sent through two LPFs in series. The output from the first filter in the chain falls off at approximately 12 dB per octave above the cut-off low-pass frequency, thus simulating the normal voicing signal depicted in Figure 7.6. The cut-off frequency for this filter is normally set to an octave above the value of the fundamental frequency. Next, the output of the second LPF should produce a smoother quasi-sinusoidal voicing if its cut-off frequency is set to approximately one octave above the value of the cut-off of the first. Some degree of quasi-sinusoidal voicing can be added to the normal voicing source, in combination with aspiration noise (see below), in order to produce the above-mentioned breathy sound quality.

The pcmusic code for this voicing mechanism is given as follows:

```
osc b20 AMP PER f1 d;            // b20=envelope voicing
blp b30 b20 b16 b16 30 d d;      // b30=pulse
NRES(b31, b30, 0dB, 0, BNV);     // b31=LPFed normal
                                 // voicing
NRES(b32, b31, –6 dB, 0, BQS);   // b32=LPFed quasi-
                                 // sinusoidal
adn b33 b31 b32;                 // b33=voice source
```

Figure 7.7 The noise source
mechanism

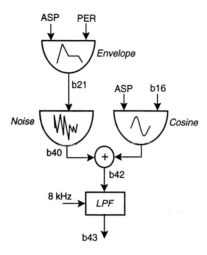

The noise source

Noise in the human voice corresponds to the generation of
turbulence by the airflow past a constriction and/or past a
relatively wide separation of the vocal folds. The resulting noise
is called *aspiration* if the constriction is located at the level of the
vocal folds; i.e. the larynx. If the constriction is located above the
larynx, the resulting noise is referred to as *fricative*. If it is
produced by a relatively wide separation of the vocal folds then
the resulting noise is called *bypass*.

Our instrument produces noise using a random number generator, an offset modulator, and a LPF (Figure 7.7). It generates
only aspiration noise; fricative and bypass are left out for the
sake of the simplicity of this study. The pcmusic code for the
noise source is given below:

```
osc b21 ASP PER f1 d;              // b21=envelope
                                   // aspiration

white b40 b21;
osc b41 ASP b16 f3 d;              // noise modulator
adn b42 b40 b41;
NRES(b43, b42, 0 dB, 0, 22624 Hz); // b43=noise source
```

7.2.2 The resonator component

On its journey through the vocal tract the sound is transformed.
Components which are close to one of the resonance frequencies
of the tract are transmitted with high amplitude, while those

Figure 7.8 The resonator component of the example instrument is composed of six band-pass filters

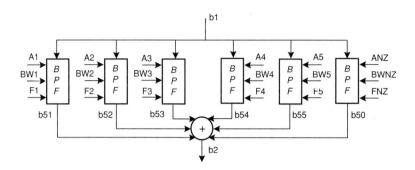

which lie far from a resonance frequency are suppressed. Much of the art of the singer lies in shaping the vocal tract in such a way that the crude source output is transformed into a desired sound. The vocal tract can be thought of as a pipe from the vocal folds to the lips plus a side-branch leading to the nose, with a cross-section area which changes considerably.

The length and shape of a particular human vocal tract determine the resonance in the spectrum of the voice signal. The length of the human vocal pipe is typically about 17 cm, which can be slightly varied by rising or lowering the larynx and shaping the lips. The cross-sectional area of the pipe is determined by the placement of the lips, jaw, tongue and velum. For the most part, however, the resonance in the vocal tract is tuned by changing its cross-sectional area at one or more points. A variety of sounds may be obtained by adjusting the shape of the vocal tract during phonation.

Five band-pass filters are appropriate for simulating a vocal tract with a length of about 17 cm (Figure 7.8). A typical female's vocal tract is 15–20 per cent shorter than that of a typical male. In this case, four filters only are enough to simulate a female's tract. Each filter introduces a peak in the magnitude spectra determined by the passband centre frequency and by the formant bandwidth. The passband centre frequencies and bandwidths of the lowest three formants vary substantially with changes in articulation, whereas the fourth and fifth formants do not vary as much.

When singing vowels, the connecting passage between the throat and the nasal cavity is closed by raising the back and soft palate. Opening this passage while singing produces a sound usually described as *singing through the nose*. The resonance of the nasal cavity is of great importance for the production of some consonants. The side-branch leading to the nasal cavity may be roughly simulated by introducing a 'nasal' sixth BPF.

The pcmusic code for the resonator component is given as follows:

```
adn b1 b33 b43;                 // b1=voice+noise
NRES(b50, b1, ANZ, FNZ, BWNZ);  // b50=nasal formant
NRES(b51, b1, A1, F1, BW1);     // b51=formant 1
NRES(b52, b1, A2, F2, BW2);     // b52=formant 2, etc.
NRES(b53, b1, A3, F3, BW3);
NRES(b54, b1, A4, F4, BW4);
NRES(b55, b1, A5, F5, BW5);
adn b2 b51 b50 b52 b53 b54 b55; // b2=whole formant
```

7.2.3 The full instrument and its pcmusic programming code

Figure 7.9 depicts the architecture of our instrument and its pcmusic score is given below; this instrument is available on the accompanying CD-ROM, in folder *Chapt7*, within pcmusic's materials. Note the extensive use of the #*define* instruction to create symbolic labels for the various synthesis parameters of the

Figure 7.9 The architecture of the instrument

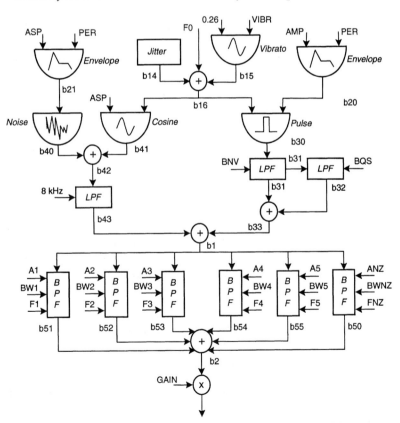

instrument, some of which represent variables whose values are provided in the note list (e.g. F0, AMP, etc.), whereas others represent constant values (F1, BW1, A1, etc.). The resonators, for example, are set with constant values to produce the vowel /a/, as in the word 'car' in English. (The values to produce the formants of other vowels can be found in Appendix 2.)

Also, the #*define* instruction is used to define the macro SING, which is a shorthand to facilitate the specification of note lists. Each SING statement needs six parameters: start time (in seconds), duration (in seconds), voicing source attenuation (in dB), aspiration noise attenuation (in dB), pitch (in Hz) and vibrato rate (in Hz), respectively.

```
//────────────────────────────────────────
#include <cmusic.h>
#include <waves.h>
//────────────────────────────────────────
#define F0      p7          // fundamental frequency
#define AMP     p5          // amplitude voicing
#define ASP     p6/2        // amplitude noise
#define VIBR    p8          // vibrato rate
#define PER     1/p4Hz      // period for envelope
                            // oscillator
#define BNV     (p7*2)*1.414  // LPF voice source -> normal
                              // voicing
#define BQS     (p7*4)*1.414  // LPF normal voicing ->
                              // quasi sinusoidal
#define F1      622.25 Hz   // formant 1 centre frequency
#define BW1     60 Hz       // formant 1 bandwidth
#define A1      0 dB        // formant 1 attenuation
#define F2      1046.5 Hz   // formant 2 centre frequency
#define BW2     70 Hz       // formant 2 bandwidth
#define A2      -7 dB       // formant 2 attenuation
#define F3      2489 Hz     // formant 3 centre frequency
#define BW3     110 Hz      // formant 3 bandwidth
#define A3      -9 dB       // formant 3 attenuation
#define F4      2900 Hz     // formant 4 centre frequency
#define BW4     130 Hz      // formant 4 bandwidth
#define A4      -12 dB      // formant 4 attenuation
#define F5      3250 Hz     // formant 5 centre frequency
#define BW5     140 Hz      // formant 5 bandwidth
#define A5      -22 dB      // formant 5 attenuation
#define FNZ     311.12 Hz   // nasal formant centre
                            // frequency
#define BWNZ 70 Hz          // nasal formant bandwidth
```

```
#define ANZ  –22 dB         // nasal formant attenuation
#define GAIN 10              // output gain

ins 0 voice;
//———Voice source———————————————————————


//———Jitter————————————————————————————
    ran b10 0.02 0.05Hz d d d;
    ran b11 0.02 0.111Hz d d d;
    ran b12 0.02 1.219Hz d d d;
    adn b13 b10 b11 b12;
    mult b14 b13 F0;              // b14=jitter factor

//———Vibrato———————————————————————————
    osc b15 0.26 VIBR f2 d;       // b15=vibrato
    adn b16 b15 b14 F0;          // b16=freq+jitter+vibrato

//———Voicing———————————————————————————
    osc b20 AMP PER f1 d;         // b20=envelope voicing
    blp b30 b20 b16 b16 30 d d;   // b30=pulse
    NRES(b31, b30, 0 dB, 0, BNV); // b31=LPFed normal
                                  // voicing
    NRES(b32, b31, –6 dB, 0, BQS); // b32=LPFed quasi-
                                   //sinusoidal
    adn b33 b31 b32;             // b33=voice source


//———Noise source——————————————————————
    osc b21 ASP PER f1 d;         // b21=envelope
                                  //aspiration

    white b40 b21;
    osc b41 ASP b16 f3 d;         // noise modulator
    adn b42 b40 b41;
    NRES(b43, b42, 0 dB, 0, 22624 Hz);
                                  // b43=noise source


//———Resonators————————————————————————
    adn b1 b33 b43;               // b1=voice+noise
    NRES(b50, b1, ANZ, FNZ, BWNZ);
                                  // b50=nasal formant
    NRES(b51, b1, A1, F1, BW1);   // b51=formant 1
    NRES(b52, b1, A2, F2, BW2);   // b52=formant 2, etc.
    NRES(b53, b1, A3, F3, BW3);
    NRES(b54, b1, A4, F4, BW4);
    NRES(b55, b1, A5, F5, BW5);
    adn b2 b51 b50 b52 b53 b54 b55;
                                  // b2=whole formant
```

```
//——Output——————————————————————
    mult b2 b2 10;
    out b2;
end;

GEN4(f1) 0, 0 –5 0.2, 1 1 0.8, 0.7 –5 1, 0;
SINE(f2);
COS(f3);

#define SING(st,dur,amp,aspir,freq,vibr)\
note st voice dur amp aspir freq vibr

SING(0.00, 0.55, –2 dB, –28 dB, 196.00 Hz, 5.2 Hz);
SING(0.55, 0.55, –4 dB, –30 dB, 155.56 Hz, 5.2 Hz);
SING(1.12, 3.79, –6 dB, –33 dB, 138.81 Hz, 5.4 Hz);
SING(5.05, 0.50, –3 dB, –31 dB, 174.61 Hz, 5.3 Hz);
SING(5.55, 2.00, 0 dB, –32 dB, 196.00 Hz, 5.3 Hz);
SING(8.02, 1.98, –3 dB, –30 dB, 155.56 Hz, 5.2 Hz);
SING(10.00, 0.60, –3 dB, –30 dB, 138.81 Hz, 5.3 Hz);
SING(10.62, 2.60, 0 dB, –32 dB, 174.61 Hz, 5.3 Hz);
ter;
```

8 Towards the cutting edge

8.1 Artificial intelligence sound synthesis

The sound domain of Western music is no longer demarcated by the boundaries of traditional acoustic instruments. Nowadays, composers have the opportunity to create music with an infinite variety of sounds, ranging from those produced by acoustic devices and different types of mechanical excitation to electronically synthesised sounds.

It is evident from the previous chapters that computer technology offers the most detailed control of the internal parameters of synthesised sounds, which enables composers to become more ambitious in their quest for more sophisticated and expressive electronic sonorities. In this case, however, the task of sound synthesis becomes more complex. A composer can set the parameters for the production of an immeasurable variety of sounds, but this is often accomplished by feeding a synthesis algorithm with streams of numerical data, either by means of an external controller (e.g. a MIDI controller) or via a score file (e.g. as in CLM). Even if the composer knows the role played by each single parameter for synthesising a sound, it is both very difficult and tedious to ascertain which values will synthesise the sound he or she wants to produce. In such a situation, higher processes of inventive creativity and abstraction become subsidiary to time-consuming, non-musical tasks.

Recent studies in acoustics, psychoacoustics and the psychology of music have vastly expanded our knowledge of the nature and perception of sounds and music. It seems, however, that this interdisciplinary knowledge we have about the nature and perception of sounds has not yet been taken into account by sound synthesis software technology. Better systems may certainly be provided if engineers devise ways for including this knowledge in a sound synthesis software. One approach to this challenge is to combine sound synthesis engineering with artificial intelligence (AI) techniques.

This chapter introduces ongoing research work being conducted by the author. We are investigating methods to combine synthesis engineering with AI techniques in order to provide sound synthesis systems with the interdisciplinary knowledge mentioned above. In the quest for more user-friendly sound synthesis programs we have designed a case study system called Artist (short for Artificial Intelligence Synthesis Tool). Artist takes a radically different route from the majority of programs currently available. Whilst most programs rely on the use of sophisticated graphic interfacing to facilitate its operation, Artist concentrates on the use of natural language; for example, the user may type or speak commands such as 'play a sound with high acuteness' or 'play a dull low-pitched sound' (Miranda, 1995b)

We do not claim that Artist's operational method is a replacement for graphic interfacing. On the contrary, we believe that both methods are complementary. Nevertheless, Artist's approach may be more suitable for a number of other applications where graphic interfacing alone does not suffice; for example, for people with impaired visual, neurological or muscular abilities.

8.1.1 The quest for artificial intelligence

One of the main objectives of AI is to gain a better understanding of intelligence but not necessarily by studying the inner functioning of the brain. The methodology of AI research is largely based upon logic, mathematical models and computer simulations of intelligent behaviour.

Of the many disciplines engaged in gaining a better understanding of intelligence, AI is one of the few that has particular interest in testing its hypotheses in practical day-to-day situations. The obvious benefit of this aspect of AI is the development of technology to make machines simulate diverse types of intelligent behaviour. For example, as a result of AI development, computers can play chess and diagnose certain types of diseases extremely well.

It is generally stated that AI as such was 'born' in the late 1940s, when mathematicians began to investigate whether it would be possible to solve complex logical problems by performing sequences of simple logical operations automatically. In fact, AI may be traced back to far before computers were available, when mechanical devices began to perform tasks previously performed only by the human mind. For example, in the 1830s, Charles Babbage, an English mathematician, invented the *Analytical Engine,* a mechanical machine that could automatically execute complex sequences of symbolic operations.

8.1.2 Introduction to Artist

Designing sounds on a computer is certainly a complex kind of intelligent behaviour. Composers engage in cognitive and physical acts in order to establish the suitability and effectiveness of their creations prior to constructing them. In attempting to solve a sound design problem, composers explore possible solutions by trying out possibilities and investigating their consequences.

When synthesising sounds to be used in a composition, composers generally have their own ideas about the possibilities of organising these sounds into a musical structure. In attempting to obtain the desired sound, the composer explores a variety of possible solutions, trying out those possibilities within his or her personal aesthetic. It is this process of exploration that has instigated our research work. This process frequently results in inconsistencies between the composer's best guess at a solution and the formulation of the problem. If no solution is found then the problem either has no solution at all or it must be redefined. Sound synthesis is seen in this context as a task which demands, on the one hand, clarification of the problem and, on the other, the provision of alternative solutions. As an example, envisage a situation in which a musician commands Artist to produce a high-pitched sound. In order to do this the system might need the expression 'high-pitched sound' to be clarified. In this case the musician would explain that 'high-pitched' means a fundamental frequency above a certain threshold. If the system still does not understand the clarification, then at least some sound should be produced (e.g. at random), which would give some chance that the sound produced might satisfy the musician's requirement. This sound may then be taken as a starting point for refinement; e.g. the musician could command the computer to 'transpose it to a high-frequency band'.

In order to aid this process, Artist focuses on the provision of the following features:

- The ability to operate the system by means of an intuitive vocabulary instead of sound synthesis numerical values. It must be said, however, that the coherence of this vocabulary depends upon several factors, ranging from how the user sets up the synthesis algorithms to the terms (or labels) of the vocabulary of sound descriptors
- The ability to customise the system according to particular needs, ranging from defining which synthesis technique will be used to the vocabulary for communication (for example, verbs and adjectives in English)
- The encouragement of the use of the computer as a collaborator in the process of exploring ideas
- The ability to aid the user in concept formation, such as the generalisation of common characteristics among sounds and their classification according to prominent sound attributes
- The ability to create contexts which exhibit some degree of uncertainty.

Currently Artist has two levels of operation: *instrument design* and *sound design*. At the instrument design level, the user customises the system by implementing a synthesis algorithm and specifying the basic terms of a vocabulary (this vocabulary may grow automatically during sound design operations). At the sound design level, the system provides facilities for the design of sounds with the synthesis algorithm and vocabulary to hand.

We suggest a method for the instrument design level which assumes that the vocabulary for sound description is intimately related to the architecture of the instrument. If one wishes to manipulate the vibrato quality of sounds, for example, it is obvious that the vibrato generator must be available for operation. As far as the sound design level is concerned, the user is required to know a few basic Artist operational commands, the capabilities of the instrument at hand and the available vocabulary for sound description.

8.1.3 Representing and processing knowledge of sound synthesis

Artist requires relatively sophisticated means to store and process data. It is not only a matter of storing synthesis parameter values on a traditional database; Artist must also store knowledge of sound synthesis as well as knowledge of the system itself (for example, the structure of the instrument, the role of each synthesis parameter and the meaning of the vocabulary in terms of synthesis parameter values, to cite but a few).

We designed the storage methods and processing mechanisms of Artist based upon a number of hypotheses on how the human mind would store and process knowledge of sound synthesis. We introduce below a selection from our hypotheses and their respective roles in Artist. Note that these hypotheses are rather pragmatic and are by no means intended to explain how the human mind works.

Hypothesis 1: the layered organisation of knowledge

Our first hypothesis is that humans tend to use a layered approach to the organisation of knowledge – including knowledge of sounds. We believe that when people think of a sound event they tend to identify its perceptual qualities and regard them as kinds of assembled perceptual units. Together, these perceptual units form a concept in the mind. This concept is part of our knowledge of the world of sound and it is connected to

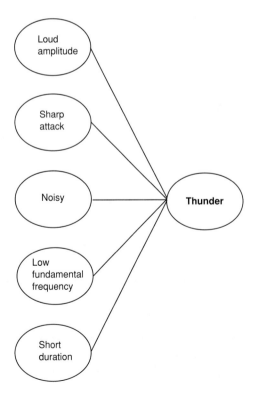

Figure 8.1 Humans tend to use a layered approach to the organisation of knowledge. When we think of a sound event, we tend to identify its perceptual qualities and regard them as kinds of assembled perceptual units

Figure 8.2 The ASS is a tree-like abstract structure consisting of *nodes*, *slots* and *links*. Nodes and slots are components and links correspond to the relations between them

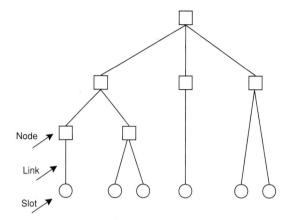

Node

Link

Slot

other concepts of the same domain through appropriate relationships.

In consequence of this hypothesis, we believe that people tend to represent information of sounds in a layered manner and use the higher-level layers to recall them. Taking as an example a thunder sound: one might associate it with attributes such as loud amplitude, sharp attack, noisy, low pitch, and medium duration. In this case, we would say that people tend to group this information and recall it simply as *thunder* instead of as *loud–sharp–noisy–low–medium* (Figure 8.1).

In order to program this hypothesis we devised a representation technique called *Abstract Sound Schema* (ASS). The ASS is a tree-like abstract structure consisting of *nodes, slots* and *links*. Nodes and slots are components and links correspond to the relations between them. Both components and relations are labelled on the ASS. Slots are grouped bottom-up into higher-level nodes, which in turn are grouped into higher-level nodes and so on until the top node (Figure 8.2).

The role of the ASS is twofold: it embodies a multilevel representation of the synthesis instrument and also provides an abstraction to represent sounds (Figure 8.3). In practice, each slot has a user-defined label and accommodates either a sound synthesis datum or a pointer towards a procedure to calculate this datum. Higher-level nodes also have user-defined labels and they correspond to the modules and sub-modules of the synthesis instrument. The top node – the root of the tree, in AI jargon – is therefore an abstraction that accommodates a sound event.

The slots must be loaded with synthesis parameter values in order to instantiate a certain sound. For each different sound produced by the instrument there is a corresponding instantia-

Figure 8.3 The ASS embodies a multilevel representation of the synthesis instrument and also provides an abstraction to represent sounds

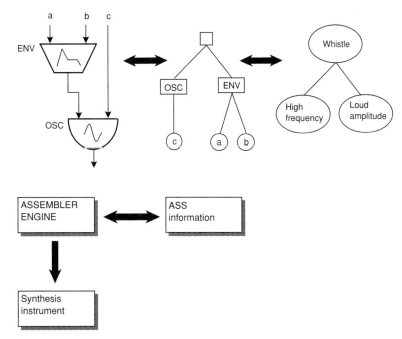

Figure 8.4 ASS information for instantiating a sound is stored in a knowledge base as a collection of slot values. An assembler engine is responsible for collecting the appropriate slot values in order to activate the synthesis instrument

tion. For example, considering the case of an instrument which simulates the vocal tract mechanism (see Chapter 7) an instantiation would correspond to the set of parameters for the simulation of a specific position of the vocal tract when producing a certain sound.

The synthesis values for instantiating sounds are recorded in a knowledge base as a collection of slot values. To produce a sound an *assembler engine* is then responsible for collecting the appropriate slot values, filling the slots of the ASS and activating the synthesis instrument (Figure 8.4).

Hypothesis 2: the identification of sound kinship

Our second hypothesis is that, in most cases, sound kinship can be defined within a multi-dimensional *attribute space*. This hypothesis is influenced by several experiments by David Ehresman (1978) and David Wessel (1979) at Ircam in Paris.

The basic idea is illustrated in Figure 8.5. If, for example, we form a three-dimensional space from three attributes, so that each attribute varies continuously; given sounds A, B and C, whose three attributes correspond to coordinates of this space, there is a sound D, such that A, B, C and D constitute the vertices of a parallelogram. In this context, sound D is expected to be

Figure 8.5 A three-dimensional space for the definition of sound kinship

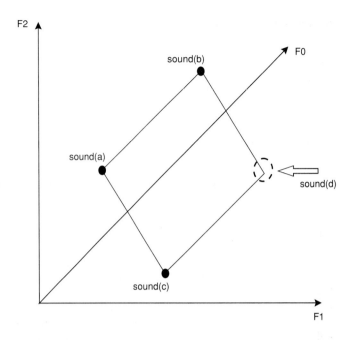

perceptually akin to sounds A, B and C. To illustrate this, suppose a configuration in which sounds are described by three attributes corresponding to the axes *x*, *y* and *z*: first formant centre frequency (represented as F1), second formant centre frequency (F2) and fundamental frequency (F0), respectively. On the one hand, *sound(a)* and *sound(b)* may have similar values for axes *x* and *y*, that is, F1 = 290 Hz and F2 = 1870 Hz, whilst, on the other hand, *sound(a)* and *sound(c)* have the same value for axis *z*, namely F0 = 220 Hz, and the value of F0 for *sound(b)* is equal to 440 Hz. If one makes a two-sound pattern – *sound(a)* followed *sound(b)* – and wishes to make an analogous pattern beginning on *sound(c)*, then according to our hypothesis the best choice would be the sound whose attributes best complete a parallelogram in the space, that is, *sound(d)*. In perceptual terms, what has varied is the attribute pitch: *sound(b)* sounds higher than *sound(a)* but preserved the same formant structure. Therefore in order to obtain an analogous two-sound pattern beginning on *sound(c)* (say, F1 = 650 Hz and F2 = 1028 Hz), the best solution would be *sound(d)*, with F1 and F2 inherited from *sound(c)* but *f0* inherited from *sound(b)*.

Although the kinship hypothesis may work well only for optimal cases (for example, it might not suit larger dimensions), we find it encouraging that a systematic description of sound events is both possible and meaningful. Artist takes advantage of the kinship hypothesis by recording the clusters of slot values

Figure 8.6 The link *akindof* is a structural link that allows for the specification of inheritance relationships in the knowledge base

Knowledge base

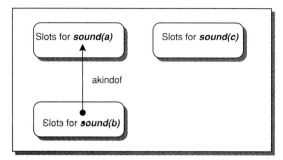

hierarchically in the knowledge base. This ability to hierarchically represent the relationship between slot collections is useful for the *inheritance mechanism* described above. This hierarchical organisation is accomplished by means of a structural link referred to as *akindof* (Figure 8.6). When a slot collection for a sound is related, by means of this link, to another slot collection at a higher level, the former inherits properties of the latter. Note in Figure 8.6 that *sound(b)* is said to be *akindof sound(a)*. This means that slots which may not eventually be defined for *sound(b)* will be instantiated with slot values taken from *sound(a)*. In practice, the assembler engine is programmed to consider that the missing slots in one level are inherited from a higher level.

Hypothesis 3: the generalisation of perceptual attributes

Finally, our third hypothesis states that when people listen to several distinct sound events, they tend to characterise them by selecting certain sound attributes which they think are important. We hypothesise that when listening to several distinct sound events, people prioritise the selection of certain attributes which are more important in order to distinguish between them. We say that in this case humans tend to make a generalisation of the perceptual attributes.

If one carefully listens to a sound event, there will probably be a large number of possible intuitive generalisations. It is therefore crucial to select those generalisations we believe to be appropriate. These depend upon several factors such as context, sound complexity, duration of events, sequence of exposure and repetition, which make a great variety of combinations possible. Humans are, however, able to make generalisations very quickly; perhaps because we never evaluate all the possibilities. We tend to limit our field of exploration and resort to a certain evaluation method. We believe that this plays an important role in imagination and memory when designing sounds and composing with them.

Artist uses inductive machine-learning techniques in order to forge this third hypothesis. These techniques are widely used in AI systems to induce general concept descriptions of objects from a set of examples (Luger and Stubblefield, 1989). In Artist we use these techniques to infer which attributes are more distinctive in a sound. The term 'more distinctive' in this case does not necessarily refer to what humans would perceive to be the most distinctive attribute of a sound. Current machine-learning techniques are not yet able to mimic all the types of strategies used by humans. Nevertheless, we propose that certain kinds of human strategies might use information theory to make generalisations. These type of strategies can be programmed on a computer using machine-learning techniques.

Artist therefore employs machine-learning algorithms that use information theory to automatically 'induce' which attributes are more distinctive on sounds given in a training set (Miranda, 1997). In this case, the training set is normally gathered by Artist itself from its own knowledge base; that is, from its collection of slot values. The result of learning is the description of sounds (or classes of sounds) in the form of rules. For example, a rule for a type of sound called *open vowel* could be:

open vowel: vibrato = normal,
            formant resonators = vowel/a/
            gender = male

            or

            vibrato = low
            formant resonators = vowel/e/
            gender = female

The interpretation of the above rule is: a sound event is an open vowel if it has normal vibrato rate, its spectral envelope corresponds to the resonance of a vowel/a/, and it is a male-like vocal sound; or it has a low vibrato rate, its spectral envelope corresponds to the resonance of a vowel/e/, and it is a female-like vocal sound.

The aim of inducing such rules is to allow the user to explore further alternatives when designing particular sounds. The user could ask Artist, for example, to 'play something that sounds similar to an open vowel'. In this case Artist would infer from the above rule which attributes are imperative to synthesise this type of sound.

8.1.4 Describing sounds by means of their attributes

It is important to emphasise that we distinguish between timbre and spectrum. The latter characterises the physical structure of a sound and it normally uses mathematics and scientific jargon for description. The former denotes the perception of these properties, and the vocabulary for its description can be highly subjective. The term 'attribute' is used in the context of Artist to refer to perceived characteristics of the timbre of a sound.

The ASS provides a good framework for sound description, but it does not, however, prescribe a methodology for describing sounds coherently. Much research is still needed on this front. We identified a variety of approaches systematically to describe sounds from their attributes (Bismark, 1974; Terhardt, 1974; Schaeffer, 1966; Slawson, 1985). It is beyond the scope of this book to survey all of these, so we have selected one approach for discussion: the *functional approach*. Here one specifies a vocabulary of sound descriptors by using the architecture of the instrument as a point of departure. The expression 'vocabulary of sound descriptors' refers to a set of user-defined labels that are used to denote a combination, or group, of synthesis parameter values. Note that the ASS is well suited for the functional approach.

To illustrate the functional approach to sound description, consider the instrument in Figure 8.7; the reader is invited to refer to Chapter 7 for a detailed description of this instrument. The voicing source model of this instrument is depicted in

Figure 8.7 The modules of a subtractive synthesis instrument

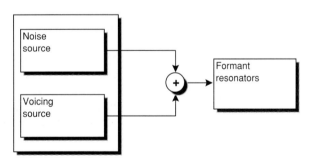

Figure 8.8 The voicing source module of the instrument depicted in Figure 8.7

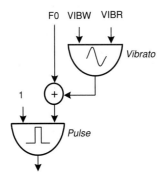

Figure 8.9 The ASS representation of the voicing source module shown in Figure 8.8

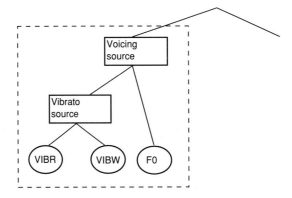

Figure 8.8 and the ASS representation of this module is illustrated in Figure 8.9. Here one could create an attribute labelled as 'voicing source' and ascribe descriptive values to it according to certain rules. For instance: *voicing source = steady low* if F0 = 80 Hz, VIBR = 5.2 Hz and VIBW = 3.5 per cent of F0 (VIBR and VIBW stand for vibrato rate and vibrato width, respectively).

Taking as another example, let us attach an attribute labelled as 'openness' to the first formant resonator of the instrument; assume that the resonator module of the instrument portrayed in Figure 8.7 is a bank of band-pass filters (Figure 8.10). In this case, the higher the value of the frequency of the first formant, the greater the openness of the sound. For example, one could define three perceptual stages, each corresponding to a certain first formant centre frequency range of values:

- *openness = low* if F1 =< 380 Hz
- *openness = medium* if F1 > 380 Hz and F1 < 600 Hz
- *openness = high* if F1 >= 600 Hz

Note that we distinguish only between two types of terms in our vocabulary for sound description: *attribute* and *attribute values*.

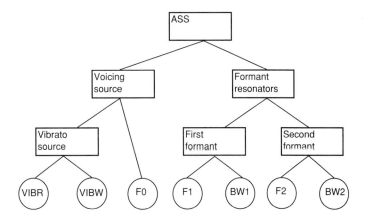

Attribute is the label we attach to a module of the instrument whereas attribute values are the labels which we give to the various sorts of output of this module, according to the variations in its synthesis parameter values. In the above example, whilst the attribute *openness* is given to the first format resonator of the instrument, the adjectives *low*, *medium* and *high* are values for this attribute.

8.1.5 The functioning of Artist

To summarise, a sound event in Artist is composed of several sound attributes. Each sound attribute in turn corresponds to one or a set of synthesis parameters. An assembler engine is responsible for computing the necessary synthesis parameter values in order to produce a sound event. The input for the assembler engine can be either the name of the desired sound event, or a set of attribute values; alternatively, one can also input numerical synthesis parameter values. The output from the assembler engine is a set of synthesis parameter values which in turn are used for synthesising the required sound event (Figure 8.11).

The assembler engine computes all the necessary synthesis parameter values for synthesising a sound event. Since the input requirement is often ill-defined (for example, an incomplete list of sound attribute values), the assembler engine is programmed to deduce the missing information by making analogies with other sound events that have similar constituents. To do this, the system automatically induces rules of prominent sound features that will identify which sound attributes are more important for describing sound events. Furthermore, given that synthesis parameter values may also be given as part of the requirement,

Figure 8.11 The functioning of Artist. The assembler engine deduces the synthesis parameters values for synthesising a sound event

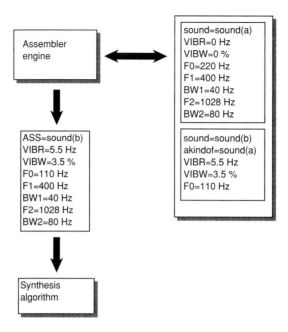

the system also handles values which may not match any available sound attribute. Likewise, the system automatically adds this new information to the knowledge base.

As an example, suppose that Artist has the following information in its knowledge:

- Sound events = { sound(a), sound(b) }
- Sound attributes = { openness = { low, high }, acuteness = { low, high }, fundamental frequency = { low, medium, high } }
- Synthesis parameters = { F1 = { 290 Hz, 400 Hz, 650 Hz }, F2 = {1028 Hz, 1700 Hz, 1870 Hz}, F0 = {220 Hz, 440 Hz, 880 Hz} }

The symbols in brackets denote possible values for the element on the left-hand side of the equation. For example, a sound attribute may value *openness, acuteness,* or *fundamental frequency.* The sound attribute *openness* may in turn value either *low* or *high.*

The assembler engine is programmed to infer from the system's ASS that, for example, *sound(a)* is described by having three attributes and values: *low openness, high acuteness* and *low fundamental frequency.* Furthermore, it can infer that *low openness* and *high acuteness* mean F1 = 290 Hz and F2 = 1870 Hz, respectively and that *low fundamental frequency* means F0 = 220 Hz.

Suppose that one inputs a request to synthesise a sound but only the information *low fundamental frequency* is given. In this case, Artist considers this input as an ill-defined requirement because the two other attributes needed to describe a sound fully are missing; in other words, there are no values for openness and acuteness. In this case, the assembler engine must allocate values for the missing attributes. This is accomplished by either consulting the induced rules of prominent sound features (the ones that identify which sound attributes are more relevant for describing a certain class of sound events) or by allocating values by chance. For this example, let us suppose that Artist synthesises *sound(a)* because it knows some rules suggesting that this sound best matches the description *low fundamental frequency*.

The current version of Artist only works effectively on large computers. Hence the reason it has not been provided on the accompanying CD-ROM. We hope to make it available to electronic musicians in the near future, either when the provision of high-performance computing via the global computer network becomes sustainable or when the power of personal computers meets the demands of the system.

8.2 Parallel computing and sound synthesis

Since computers were invented there have been no major changes in their basic design. The main core of most currently available computers is a central processing unit (CPU), provided with a single memory component, which executes instructions sequentially. This configuration is commonly referred to as *von Neumann architecture* (after its inventor), but for the purposes of this book we use the more generic acronym SISD, for Single Instruction Single Data.

SISD machines call for the concept of sequential programming in which a program consists of a list of instructions to be executed one after the other. Although such machines may now work so fast that they appear to be executing more than one instruction at a time, the sequential programming paradigm still remains for most techniques and tools for software development, including sound synthesis.

The sequential paradigm works well for most of the ordinary tasks for which we need the computer but scientists are very aware of the limitations of current computer technology. For example, if computers were programmed to display intelligent or customary behaviour the sequential paradigm would fail to model many aspects of human intelligence and natural systems. Particular mental processes seem to be better modelled as a system distributed over thousands of processing units as, for example, an idealised model of brain tissue. It is possible to simulate this distributed type of behaviour on fast SISD machines by using a limited number of processing units, but no processor can support the number of units required to study the behaviour of the more sophisticated distributed machine models.

Although the speed of SISD machines has increased substantially over the last few years, these improvements are beginning to reach the limits of physics, beyond which any further development would be impossible. In particular, the speed of light is a constant upper boundary for the speed of signals within a computer, and this speed is already proving too slow to allow further increases in the speed of processors. The only possible way to meet the demand for increasing performance is to abandon dependence upon single-processor hardware and look towards parallel processing.

As far as sound synthesis is concerned, parallel computing is commonly associated with the possibility of breaking the speed limits imposed by SISD machines. However much faster sequential computers become, parallel computing will still offer interesting programming paradigms for the development of new synthesis techniques. From a composer's perspective, it is this latter aspect of parallel computing that excites interest.

8.2.1 Architectures for parallel computing

Two main approaches are used to build parallel computers: SIMD (for Single Instruction Multiple Data) and MIMD (for Multiple Instructions Multiple Data). Whilst the former employs many processors simultaneously to execute the same program on different data, the latter employs several processors to execute different instructions on different data. Both approaches underpin different programming paradigms and have their own merits and inadequacies.

The SIMD architecture

SIMD-based computers employ a large amount of interconnected *processing elements* (PE) running the same programming instructions concurrently but working with different data (Figure 8.12). Each PE has its own local processing memory, but the nature of the PEs and the mode of communication between them varies for different implementations. PEs are usually simpler than conventional processing units used on SISD machines because PEs do not usually require individual instruction fetch mechanisms; this is managed by a master control unit (MCU). Communication between PEs commonly involves an

Figure 8.12 The SIMD architecture employs a large amount of interconnected processing elements (PE) running the same programming instructions concurrently but working on different data

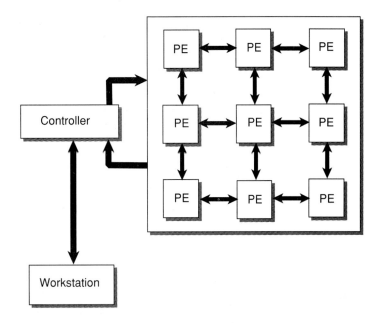

orthogonal grid of data pathways and each PE can often communicate directly with its four or eight nearest neighbours.

SIMD computers are essentially synchronous models because all PEs execute the same instruction at the same time. The only control the programmer has over an individual PE is to allow or prevent it from executing the instruction. This makes SIMD computers easier to program than MIMD computers (see below) because the control of different programs running on each processor of a MIMD machine becomes very demanding as the number of processors increases. SIMD architectures therefore suit the provision of very large arrays of PEs; current SIMD computers may have as many as 65 536 PEs.

The MIMD architecture

MIMD-based computers employ independent processors running different programming instructions concurrently and working on different data. There are a number of different ways to construct MIMD machines, according to the relationship of processors to memory and to the topology of the processors.

Regarding the relationship between processor and memory, we identify two fundamental types: *shared global memory* and *private local memory*. In shared global memory, each processor has a direct link to a single global memory via a common pathway, or *bus* in computer science jargon (Figure 8.13). In this case, proces-

Figure 8.13 The processors of MIMD–based computers can only communicate with each other via the global memory

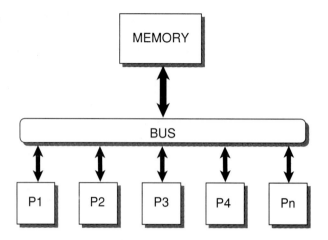

sors only communicate with each other via the global memory. This type of MIMD architecture is sometimes preferable because they are relatively straightforward to program. The major drawback is that the number of processors should be limited to accommodate the capacity of the pathway. Otherwise, too many processors competing to access the memory would need complicated mechanisms to prevent data traffic jams.

Figure 8.14 The most often used topologies of MIMD transputers are binary trees (a), two-dimensional grids (b) and hypercubes (c)

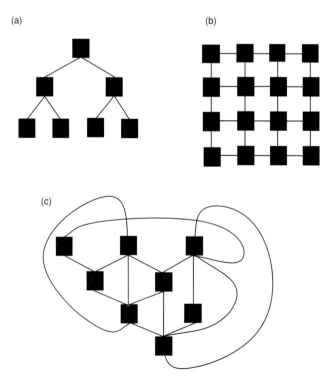

In order to employ large amounts of MIMD processors it is necessary to provide each processor with its own local memory. In this case, each unit is normally encapsulated in a microchip called a *transputer*.

Since MIMD transputers execute different instructions and work on different data, the interconnection between them must be carefully planned. It is unthinkable to connect each transputer to all other transputers. The number of connections rises as the square of the number of units. From a number of possibilities, most customarily used topologies are *binary trees, two-dimensional grids* and *hypercubes* (Figure 8.14). Sophisticated MIMD implementations allow for user-configuration of topologies to suit particular needs.

MIMD computers are essentially asynchronous because each processor or transputer may proceed at its own rate, independently of the behaviour of the others. They may, of course, perform in synchronous mode if required.

8.2.2 Parallel programming strategies

It seems likely that within a few years computers will be based upon a parallel architecture of some kind. Some of these machines will probably make their parallel architecture transparent to high-level programmers and will somehow allow for the processing of sequential programs as well. In this case, sequential programs must to be adapted either manually by a programmer or automatically by the machine itself.

Cutting-edge technology and best achievable performance will certainly require software explicitly designed for parallelism. A program that has been specifically designed for a parallel architecture will perform unquestionably better than a converted sequential program.

There are two different but closely related parallel programming strategies: *decomposition* and *parallelisation*. The former employs methods to decompose a sequential program for parallel processing and the latter uses techniques to design parallel programs from scratch. All these strategies are interrelated and very often the solution to a problem is best achieved by applying a combination of them.

Decomposition

The main objective of decomposition is to reduce the execution time of a sequential program. In general, it is possible to split a sequential program into various parts, some of which would have good potential for concurrent processing. By allocating these parts to various parallel processors, the runtime of the overall program could be drastically reduced.

Decomposition requires caution because the splitting of a program into a large number of processors incurs considerable costs. For example, it should be estimated that individually processed parts might require demanding communication mechanisms to exchange data between them. That is, a large amount of processors does not necessarily lead to good performance.

We identify three main strategies for decomposition: *trivial, pipeline* and *geometric*. A trivial decomposition takes place when the original sequential program processes a great amount of data with no interdependencies. In this case, there are no major technical problems to split the task, since parallelism may be achieved simply by running copies of the entire program on various processors concurrently. Conversely, a pipeline decomposition strategy is applied when a program can be split into various modules and allocated to different processors. All input data still passes through each of the modules sequentially but as they move through the pipeline the modules work simultaneously (Figure 8.15). The first input element passes through the first stage of the pipeline and, after being processed, it then passes on to the next stage of the pipeline. Whilst this first element is being processed at the second stage, the second input element is fed into the first stage for processing, and so on. All stages work simultaneously because each has its own exclusive processor. Finally, geometric decomposition splits the data rather than the program. This strategy is suited for those problems where identical operations are applied to different parts of a large data set. As with the case of trivial decomposition, each processor has a copy of the whole program. The difference between trivial and geometric decomposition is that the latter is designed to cope with data dependencies; that is, the result of the operation of one data subset may require information about the operation of other data subsets. Geometric decomposition works best when the data subsets need to interact with their neighbouring subsets. In this

Figure 8.15 In pipeline decomposition, each input data passes through each of the modules sequentially but as they move through the pipeline the modules work simultaneously

case the data is distributed to a certain number of processors arranged in a lattice and the program will have to be provided with mechanisms for passing messages across the processors.

Parallelisation

There are two main different approaches to the design of a parallel program: *specialist processing* and *task farm processing*. These approaches have many links to the decomposition categories introduced above and both concepts certainly complement each other.

Specialist processing

Specialist processing achieves parallelism by dividing the problem into specific types of tasks. For example, if the problem is to build a piano, then there will be a variety of specialised tasks such as making the keys, strings, case, etc. Once all pieces are complete, the piano is then assembled. Specialist workers are assigned for each task in order to ensure the quality of the components and increase the speed of construction. However, if the objective is to build only one piano, workers may not be able to start working simultaneously because some will have to wait for other colleagues to finish in order to start their job. Conversely, if the aim is to manufacture several pianos, then it may be possible to keep all workers occupied for most of the time by 'pipelining' the tasks (see *pipeline decomposition* above). In any case, it is essential to establish efficient channels of communication between the workers so that they can coordinate their work.

Task farm processing

In task farm processing, the problem is divided into several tasks but not necessarily targeted for specialist processors. To continue the above-mentioned metaphor of the piano, there are no specialist workers in the team here because it is assumed that all of them are capable of carrying out any of the tasks for building a piano. In this case, there is a master who holds a list of tasks to be performed. Each worker takes a task and when it is accomplished he or she comes back and selects another. Communication occurs mostly between a worker and the master; for the sake of maximum performance, workers are not encouraged to talk to each other because their attention must not be diverted from the main task.

8.2.3 Benefits for sound synthesis

In several ways, parallel computing concepts are making their way into the synthesis technology, from the design of dedicated VLSI (Very Large Scale Integration) chips to the implementation of new paradigms for sound synthesis.

Signal processing level

Dedicated digital signal processors (called as DSP) with multiple functional units and pipeline methods have been encapsulated into VLSI chips and today are commonly found in a variety of hardware for sound synthesis. The functioning of a DSP is driven by a set of instructions that is loaded into its own memory from a host computer. The DSP then cycles continuously through these instructions until it receives a message to halt; a sample is produced at each cycle of the DSP. Special DSP arrangements may form arrays of DSP for simultaneously processing blocks of samples. In this case, the output for each cycle is an array of samples. Also, parallel configurations of specially designed general purpose microprocessors based on RISC (Reduced Instruction Set Computer) technology have been used on several occasions.

Software synthesis programming level

In software synthesis programming there have been a few attempts to decompose existing synthesis programming languages for concurrent processing. Remarkable results have been reported by the composer Peter Manning and his collaborators at Durham University, who have managed to run Csound concurrently on a parallel machine (Bailey *et al.*, 1990). Like pcmusic and CLM, Csound is a synthesis programming language in which instruments are designed by connecting many different synthesis units. One or more instruments are saved in a file called the *orchestra*. Then the parameters to control the instruments of the orchestra are specified on an associated file called the *score* (see Chapter 1). This score file is organised in such a way that each line corresponds to a stream of parameters to produce a sound event (or note) on one of the instruments of the orchestra. For instance:

*i*1 0 1.0 440 90
*i*2 1 0.5 220 80

In the example above, each line specifies five parameters for two different instruments, including the 'names' of the instruments (i.e. *i*1 and *i*2, respectively). Each line produces a note.

The parallel implementation of the Durham team is inspired by the task farm paradigm discussed earlier. An copy of the entire orchestra is distributed to each processor and the score is considered as a list of tasks for the processors; that is, each processor selects a line from the score for processing.

Synthesis modelling level

At this level a parallel computing paradigm is normally embedded in the synthesis model itself. As an example we cite Chaosynth, a synthesis system developed by the author in collaboration with the engineers of the Edinburgh Parallel Computing Centre, in Scotland (Miranda, 1995a).

Briefly, Chaosynth uses cellular automata (CA) to control the parameters of a granular synthesis instrument. The granular synthesis of sounds involves the production of sequences of thousands of short sonic particles (for example, 35 milliseconds) in order to form larger sound events (see Chapter 5).

CA are mathematical models of dynamic systems in which space and time are discrete and quantities take on a finite set of discrete values, and are often represented as a regular array with a discrete variable at each site, referred to as a cell. The state of the CA is specified by the values of the variables at each cell. It evolves in synchronisation with the tick of an imaginary clock, according to a global function that determines the value of a cell based upon the value of its neighbourhood. As implemented on a computer, the cells are represented as a grid of tiny rectangles, whose states are indicated by different colours.

The algorithm of the CA used in Chaosynth is fully explained in the documentation included on the CD-ROM that accompanies this book. Each grain produced by Chaosynth contains a number of partials produced by a bank of oscillators, as with additive synthesis (see Chapter 6). The frequencies of the partials are determined according to the evolution of the CA.

In Chaosynth, the states of the CA cells represent frequency values rather than colours. The oscillators are associated to a group of cells, and in this case the frequency for each oscillator is established by the arithmetic mean of the frequencies of the cells associated to the respective oscillator. At each cycle of the CA, Chaosynth produces one grain and as the CA evolves, the components of each grain change therein. The size of the grid, the number of oscillators and other parameters for the CA are all configurable by the user.

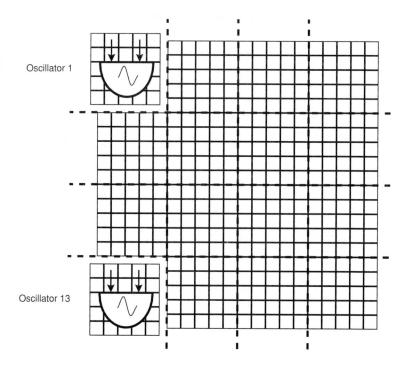

Figure 8.16 In Chaosynth, the CA grid is subdivided into a number of contiguous rectangular portions in such a way that all of them are handled concurrently by different processors. In this case, the oscillators are computed in parallel

CA are intrinsically suited for parallel processing. They are arrangements of cells, each of which is updated every cycle, according to a specific global rule that takes into account the state of the neighbouring cells. This is a typical problem for geometrical decomposition with message passing. In this case, the grid is subdivided into a number of contiguous rectangular parts in such a way that all of them are handled concurrently by different processors. The oscillators of Chaosynth are therefore computed in parallel (Figure 8.16).

On the accompanying CD-ROM is a version of Chaosynth which has been adapted by programming wizard Joe Wright to run sequentially on PC-compatible computers under Windows 95.

Appendix 1:
Mathematical specifications

Amplitude modulation

The amplitude of the modulator is specified by an amount of the offset amplitude value (a_c) in relation to a modulation index (mi):
$a_m = a_c \times mi$.

In simple AM, the spectrum of the resulting signal contains energy at thee frequencies: the frequency of the carrier (f_c) plus two sidebands ($f_c - f_m$ and $f_c + f_m$ respectively). The amplitude of the carrier frequency remains unchanged, whilst the amplitudes of the sidebands are calculated as follows: $a_c \times (0.5 \times mi)$.

Frequency modulation

List of scaling factors (not exhaustive)

i	$B_0(i)$	$B_1(i)$	$B_2(i)$	$B_3(i)$	$B_4(i)$	$B_5(i)$	$B_6(i)$	$B_7(i)$	$B_8(i)$
0.0	1.000	0.000	0.000	0.000	0.000	0.000	0.000	0.000	0.000
0.5	0.938	0.242	0.030	0.002	0.000	0.000	0.000	0.000	0.000
1.0	0.765	0.440	0.115	0.019	0.002	0.000	0.000	0.000	0.000
1.5	0.512	0.558	0.232	0.060	0.011	0.001	0.000	0.000	0.000
2.0	0.223	0.576	0.352	0.129	0.034	0.007	0.001	0.000	0.000
2.5	−0.048	0.500	0.446	0.216	0.073	0.020	0.004	0.000	0.000
3.0	−0.260	0.340	0.486	0.309	0.132	0.043	0.011	0.002	0.000
3.5	−0.380	0.137	0.458	0.386	0.204	0.080	0.025	0.006	0.001
4.0	−0.400	−0.066	0.364	0.430	0.281	0.132	0.050	0.015	0.004
4.5	−0.032	−0.231	0.217	0.424	0.348	0.194	0.084	0.030	0.009
5.0	−0.177	−0.327	0.046	0.364	0.391	0.261	0.131	0.053	0.018
5.5	−0.006	−0.341	−0.117	0.256	0.396	0.320	0.186	0.086	0.033
6.0	0.150	0.276	−0.242	0.115	0.357	0.362	0.245	0.130	0.056
6.5	0.260	−0.153	−0.307	−0.035	0.274	0.373	0.300	0.180	0.088
7.0	0.300	−0.004	−0.301	−0.167	0.157	0.347	0.340	0.233	0.128
7.5	0.266	0.135	−0.230	−0.258	0.023	0.283	0.354	0.283	0.174
8.0	0.171	0.234	−0.113	−0.291	−0.105	0.185	0.337	0.320	0.223
8.5	0.041	0.273	0.022	−0.262	−0.207	0.067	0.287	0.337	0.270
9.0	−0.090	0.245	0.145	−0.180	−0.265	−0.055	0.204	0.327	0.305
9.5	−0.194	0.161	0.228	−0.065	−0.270	−0.161	0.100	0.286	0.323
10.0	−0.245	0.043	0.254	0.058	0.220	−0.234	−0.014	0.216	0.317

Single carrier with parallel modulators example

The amplitude scaling factors for the calculation of the FM spectrum of a single carrier with two parallel modulators scheme result from the multiplication of the respective Bessel functions: $B_n(i_1) \times B_m(i_2)$. In this case, the frequencies of the spectrum is calculated as follows:

$$f_c - (n \times f_{m1}) + (m \times f_{m2})$$
$$f_c - (n \times f_{m1}) - (m \times f_{m2})$$
$$f_c + (n \times f_{m1}) + (m \times f_{m2})$$
$$f_c + (n \times f_{m1}) - (m \times f_{m2})$$

Example:

$$i_1 = 1.5$$
$$i_2 = 1.0$$
$$f_c = 440 \text{ Hz}$$
$$f_{m1} = 100 \text{ Hz}$$
$$f_{m2} = 30 \text{ Hz}$$

$B_n(1.5) \times B_m(1.0)$ for $440 - (n \times 100) + (m \times 30)$
$$440 - (n \times 100) - (m \times 30)$$
$$440 + (n \times 100) + (m \times 30)$$
$$440 + (n \times 100) - (m \times 30)$$

where $n = \{0, 1, 2\}$ and $m = \{0, 1, 2\}$.

The frequencies and amplitude scaling factors are calculated as follows (Figure A.1):

(a) $B_0(1.5) \times B_0(1.0) = 0.512 \times 0.765 = 0.40$
$440 - (0 \times 100) + (0 \times 30) = 440 \text{ Hz}$
$440 - (0 \times 100) - (0 \times 30) = 440 \text{ Hz}$
$440 + (0 \times 100) + (0 \times 30) = 440 \text{ Hz}$
$440 + (0 \times 100) - (0 \times 30) = 440 \text{ Hz}$

(b) $B_0(1.5) \times B_1(1.0) = 0.512 \times 0.440 = 0.22$
$440 - (0 \times 100) + (1 \times 30) = 470 \text{ Hz}$
$440 - (0 \times 100) - (1 \times 30) = 410 \text{ Hz}$
$440 + (0 \times 100) + (1 \times 30) = 470 \text{ Hz}$
$440 + (0 \times 100) - (1 \times 30) = 410 \text{ Hz}$

(c) $B_0(1.5) \times B_2(1.0) = 0.512 \times 0.115 = 0.06$
$440 - (0 \times 100) + (2 \times 30) = 500 \text{ Hz}$
$440 - (0 \times 100) - (2 \times 30) = 380 \text{ Hz}$
$440 + (0 \times 100) + (2 \times 30) = 500 \text{ Hz}$
$440 + (0 \times 100) - (2 \times 30) = 380 \text{ Hz}$

(d) $B_1(1.5) \times B_0(1.0) = 0.558 \times 0.765 = 0.42$
$440 - (1 \times 100) + (0 \times 30) = 340 \text{ Hz}$
$440 - (1 \times 100) - (0 \times 30) = 340 \text{ Hz}$
$440 + (1 \times 100) + (0 \times 30) = 540 \text{ Hz}$
$440 + (1 \times 100) - (0 \times 30) = 540 \text{ Hz}$

(e) $B_1(1.5) \times B_1(1.0) = 0.558 \times = 0.440 = 0.24$
$440 - (1 \times 100) + (1 \times 30) = 370 \text{ Hz}$
$440 - (1 \times 100) - (1 \times 30) = 310 \text{ Hz}$
$440 + (1 \times 100) + (1 \times 30) = 570 \text{ Hz}$
$440 + (1 \times 100) - (1 \times 30) = 510 \text{ Hz}$

(f) $B_1(1.5) \times B_2(1.0) = 0.558 \times 0.115 = 0.06$
$440 - (1 \times 100) + (2 \times 30) = 400 \text{ Hz}$
$440 - (1 \times 100) - (2 \times 30) = 280 \text{ Hz}$
$440 + (1 \times 100) + (2 \times 30) = 600 \text{ Hz}$
$440 + (1 \times 100) - (2 \times 30) = 480 \text{ Hz}$

(g) $B_2(1.5) \times B_0(1.0) = 0.232 \times 0.765 = 0.18$
    $440 - (2 \times 100) + (0 \times 30) = 240 \text{ Hz}$
    $440 - (2 \times 100) - (0 \times 30) = 240 \text{ Hz}$
    $440 + (2 \times 100) + (0 \times 30) = 640 \text{ Hz}$
    $440 + (2 \times 100) - (0 \times 30) = 640 \text{ Hz}$

(h) $B_2(1.5) \times B_1(1.0) = 0.232 \times 0.440 = 0.10$
    $440 - (2 \times 100) + (1 \times 30) = 270 \text{ Hz}$
    $440 - (2 \times 100) - (1 \times 30) = 210 \text{ Hz}$
    $440 + (2 \times 100) + (1 \times 30) = 670 \text{ Hz}$
    $440 + (2 \times 100) - (1 \times 30) = 610 \text{ Hz}$

(i) $B_2(1.5) \times B_2(1.0) = 0.232 \times 0.115 = 0.02$
    $440 - (2 \times 100) + (2 \times 30) = 300 \text{ Hz}$
    $440 - (2 \times 100) - (2 \times 30) = 180 \text{ Hz}$
    $440 + (2 \times 100) + (2 \times 30) = 700 \text{ Hz}$
    $440 + (2 \times 100) - (2 \times 30) = 580 \text{ Hz}$

Figure A.1 An example of an FM spectrum generated by a single carrier with two parallel modulators

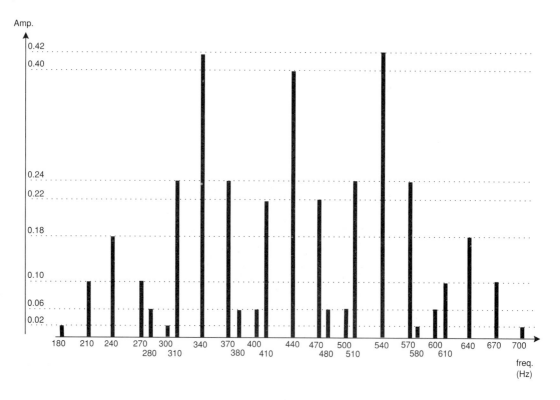

Waveshaping

To a large extent the best transfer functions for waveshaping synthesis are described using polynomials. The amplitude of the signal input to the waveshaper is represented by the variable x and the output is denoted by $F(x)$, where F is the function and d are amplitude coefficients:

$$F(x) = d_0 + d_1 x + d_2 x^2 + \ldots + d_n x^n$$

Polynomials are useful because they can predict the highest partial that will be generated by the waveshaper. If the input is a sinusoid and the transfer function a polynomial of order n, then the waveshaper produces harmonics up to the nth harmonic. In this case, the amplitudes of the harmonics can be calculated according the the table in Figure A.2. This table enumerates the amplitudes of the harmonics produced by a term in the polynomial when the amplitude of the input sinusoid is 1. It also indicates that a term does not produce more harmonics than its exponent. For example, the transfer function $F(x) = x + x^2 + x^3 + x^4$ will produce the first, second, third and fourth harmonics. The amplitudes of the harmonics are calculated as the sum of the contributions of each term:

$$a_{h0} = 0.5 + 0.375 = 0.875$$
$$a_{h1} = 1 + 0.75 = 1.75$$
$$a_{h2} = 0.5 + 0.5 = 1$$
$$a_{h3} = 0.25$$
$$a_{h4} = 0.125$$

Figure A.2 The amplitude of the harmonics produced by a term in the polynomial when the amplitude of the input sinusoid is equal to one

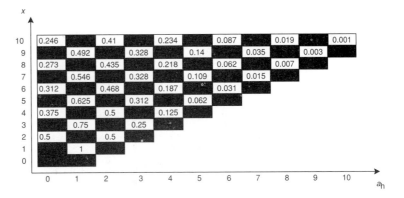

Note that the even power of x produces even harmonics, whilst the odd power produces odd ones. This affords musicians independent control of the odd and even harmonics of the sound.

When the waveform applied to the waveshaper is not a sinusoid the resulting spectrum is more difficult to predict. The calculation of output spectra for non-sinusoidal input signal is beyond the scope of this book.

Chebyshev polynomials

Chebyshev polynomials are represented as follows: $T_k(x)$ where k represents the order of the polynomial and x represents a sinusoid. Chebyshev polynomials have the useful property that when a cosine wave with amplitude equal to one is applied to $T_k(x)$, the resulting signal is a sinewave at the kth harmonic. For example, if a sinusoid of amplitude equal to one is applied to a transfer function given by the seventh-order Chebyshev polynomial, the result will be a sinusoid at seven times the frequency of the input. Chebyshev polynomials for T_1 through T_{10} are given as follows:

$$T_1(x) = x$$
$$T_2(x) = 2x^2 - 1$$
$$T_3(x) = 4x^3 - 3x$$
$$T_4(x) = 8x^4 - 8x^2 + 1$$
$$T_5(x) = 16x^5 - 20x^3 + 5x$$
$$T_6(x) = 32x^6 - 48x^4 + 18x^2 - 1$$
$$T_7(x) = 64x^7 - 112x^5 + 56x^3 - 7x$$
$$T_8(x) = 128x^8 - 256x^6 + 160x^4 - 32x^2 + 1$$
$$T_9(x) = 256x^9 - 576x^7 + 432x^5 - 120x^3 + 9x$$
$$T_{10}(x) = 512x^{10} - 1280x^8 + 1120x^6 - 400x^4 + 50x^2 - 1$$

Because each separate polynomial produces a particular harmonic of the input signal, a certain spectrum composed of various harmonics can be obtained by summing a weighted combination of Chebyshev polynomials, one for each desired harmonic. For example, the transfer function to produce a spectrum containing the first, second, third and fifth harmonics can be determined as follows:

Figure A.3 An example of a spectrum

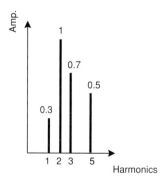

$$F(x) = w_1T_1(x) + w_2T_2(x) + w_3T_3(x) + w_5T_5(x)$$

The weighting values (w_1, w_2, w_3 and w_5) correspond to the relative amplitudes of the respective harmonic components.

As an example, consider the calculation of the tranfer function to produce the spectrum shown in Figure A.3:

$$F(x) = 0.3T_1(x) + 1T_2(x) + 0.7T_3(x) + 0.5T_5(x)$$
$$F(x) = 0.3(x) + 1(2x^2 - 1) + 0.7(4x^3 - 3x) + 0.5(16x^5 - 20x^3 + 5x)$$
$$F(x) = 0.3x + 2x^2 - 1 + 2.8x^3 - 2.1x + 8x^5 - 10x^3 + 2.5x$$
$$F(x) = 8x^5 - 7.2x^3 + 2x^2 - 0.7x - 1$$

Appendix 2:
Formant values

Vowel /a/ as in the word 'car' (in English):
Male:

F1 = 622.25 Hz	BW1= 60 Hz	A1 = 0 dB
F2 = 1046.5 Hz	BW2 = 70 Hz	A2 = –7 dB
F3 = 2489 Hz	BW3 = 110 Hz	A2 = –9 dB

Female:

F1 = 783.99 Hz	BW1 = 80 Hz	A1 = 0 dB
F2 = 1174.7 Hz	BW2 = 90 Hz	A2 = –4 dB
F3 = 2793.8 Hz	BW3 = 120 Hz	A3 = –20 dB

Vowel /e/ as in the word 'bed' (in English):
Male:

F1 = 392 Hz	BW1= 40 Hz	A1 = 0 dB
F2 = 1661.2 Hz	BW2 = 80 Hz	A2 = –12 dB
F3 = 2489 Hz	BW3 = 100 Hz	A2 = –9 dB

Female:

F1 = 392 Hz	BW1 = 60 Hz	A1 = 0 dB
F2 = 1568 Hz	BW2 = 80 Hz	A2 = –24 dB
F3 = 2793.8 Hz	BW3 = 120 Hz	A3 = –30 dB

Vowel /i/ as in the word 'it' (in English):
Male:

F1 = 261.63 Hz	BW1= 60 Hz	A1 = 0 dB
F2 = 1760 Hz	BW2 = 90 Hz	A2 = –30 dB
F3 = 2489 Hz	BW3 = 100 Hz	A2 = –16 dB

Female:
F1 = 349.23 Hz   BW1 = 50 Hz   A1 = 0 dB
F2 = 1661.2 Hz   BW2 = 100 Hz   A2 = –20 dB
F3 = 2793.8 Hz   BW3 = 120 Hz   A3 = –30 dB

Vowel /o/ as in the word 'hot' (in English):
Male:
F1 = 392 Hz   BW1= 40 Hz   A1 = 0 dB
F2 = 783.99 Hz   BW2 = 80 Hz   A2 = –11 dB
F3 = 2489 Hz   BW3 = 100 Hz   A2 = –21 dB
Female:
F1 = 440 Hz   BW1 = 70 Hz   A1 = 0 dB
F2 = 783.99 Hz   BW2 = 80 Hz   A2 = –9 dB
F3 = 2793.8 Hz   BW3 = 100 Hz   A3 = –16 dB

Vowel /u/ as in the word 'rule' (in English):
Male:
F1 = 349.23 Hz   BW1= 40 Hz   A1 = 0 dB
F2 = 587.33 Hz   BW2 = 80 Hz   A2 = –20 dB
F3 = 2489 Hz   BW3 = 100 H   A2 = –32 dB
Female:
F1 = 329.63 Hz   BW1 = 50 Hz   A1 = 0 dB
F2 = 739.99 Hz   BW2 = 60 Hz   A2 = –12 dB
F3 = 2793.8 Hz   BW3 = 170 Hz   A3 = –30 dB

References

Adrien, J.–M. (1991). The missing link: modal synthesis. In De Poli, G. *et al.* (eds), *Representation of Musical Signals*, Cambridge, MA: The MIT Press.

Arcela, A. (1994). A Linguagem SOM-A para Síntese Aditiva, *I Simpósio Brasileiro de Computação e Música*, Caxambú, Brazil, pp. 33–43.

Bailey, N., *et al.* (1990). Concurrent Csound: parallel execution for high speed direct synthesis, *Proceedings of the International Computer Music Conference*, Glasgow.

Bayle, F. (1993). *Musique Acousmatique*, Paris: INA/GRM – Edition Buchet/Chastel.

Berg, P. (1979). PILE – A language for sound synthesis, *Computer Music Journal*, **3**, 1 30–41.

Bismark, G. von (1974). Timbre of steady sounds: a factorial investigation of its verbal attributes, *Acústica*, **30**, 146–158.

Chandra, A. (1994). The linear change of waveform segments causing non-linear changes of timbral presence, *Contemporary Music Review: Timbre Composition in Electroacoustic Music*, **10**, 2, 157–169.

Chareyron, J. (1990). Digital synthesis of self-modifying waveforms by means of linear automata, *Computer Music Journal*, **14**, 4, 25–40.

Chowning, J. (1973). The synthesis of complex audio spectra by means of frequency modulation, *Journal of the Audio Engineering Society*, **21**, 7, 526–534.

Chowning, J. and Bristow, D. (1986). *FM Theory and Applications*, Tokyo: Yamaha Music Foundation.

Deutch, D. (ed.) (1982). *The Psychology of Music*, Orlando, FL: Academic Press.

Ehresman, D. and Wessel, D. (1978). *Perception of Timbral Analogies*, Rapports Ircam, Paris: Centre Georges Pompidou.

Gabor, D. (1947). Acoustical quanta and the theory of hearing, *Nature*, **159**, 591–594.

Helmholtz, H.L.F. (1885). *On the sensations of tone as a physiological basis for the theory of music*, London: Longmans, Green and Co.

Holtzman, S.R. (1978). Description of an automated digital sound synthesiser instrument, DAI Research Paper No. 59, Department of Artificial Intelligence, University of Edinburgh.

Howard, D.M. and Angus, J. (1996). *Acoustics and Psychoacoustics*, Oxford: Focal Press.

Hutchins, B. (1973). Experimental electronic music devices employing Walsh functions, *Journal of the Audio Engineering Society*, **21**, 8, 640–645.

Hutchins, B. (1975). Application of real-time Hadamard transform network to sound synthesis, *Journal of the Audio Engineering Society*, **23**, 5, 558–562.

Kaegi, W. and Tempellars, S. (1978). VOSIM – a new sound synthesis system, *Journal of the Audio Engineering Society*, **26**, 6, 418–426.

Karplus, K. and Strong, A. (1983). Digital synthesis of plucked string and drum timbres, *Computer Music Journal*, **7**, 2, 43–55.

Luger, G.F. and Stubblefield, W. A. (1989). *Artificial Intelligence and the Design of Expert Systems*, Redwood City, CA: The Benjamin/Cummings Publishing Company.

Maconie, R. (1976). *The Works of Stockhausen*, London: Marion Boyars.

Maconie, R. (1989). *Stockhausen on Music*, London: Marion Boyars.

Manning, P. (1987). *Electronic and Computer Music*, Oxford: Oxford University Press.

McAdams, S. and Deliege, I. (issue eds) (1985). *Contemporary Music Review: Music and the Cognitive Sciences*, **4**.

Miranda, E.R. (1993). Cellular automata music: an interdisciplinary project, *Journal of New Music Research (formely Interface)*, **22**, 3–21.

Miranda, E.R. (1995a). Chaosynth: Um sistema que utiliza um autômato celular para sintetizar partículas sônicas, *II Simpósio Brasileiro de Computação e Música*, Canela, Brazil, pp. 205–212.

Miranda, E.R. (1995b). An artificial intelligence approach to sound design, *Computer Music Journal*, **19**, 2, 59–75.

Miranda, E.R. (1997). Machine learning of sound attributes: a case study, *Leonardo Music Journal*, **7**, 47–54.

Moore, F.R. (1990). *Elements of Computer Music*, Englewood Cliffs, NJ: Prentice Hall.

Norman, K. (issue ed.) (1996). *Contemporary Music Review: A Poetry of Reality: Composing with Recorded Sound*, **15**, 1–2.

Roads, C. (1988). Introduction to granular synthesis, *Computer Music Journal*, **12**, 2, 11–13.

Rodet, X. (1984). Time-domain formant-wave-function synthesis, *Computer Music Journal*, **8**, 3, 9–14.

Rumsey, F. (1996). *The Audio Workstation Handbook*, Oxford: Focal Press.

Rumsey, F. (1994). *MIDI Systems and Control*, Oxford: Focal Press.

Rumsey, F. and Watkinson, J. (1995). *The Digital Interface Handbook*, 2nd edition, Oxford: Focal Press

Russ, M. (1996). *Sound Synthesis and Sampling*, Oxford: Focal Press.

Schaeffer, P. (1966). *Traité des objets musicaux*, Paris: Editions du Seuil.

Schottstaedt, B. (1994). Machine Tongues XVII: CLM: Music V meets Common Lisp, *Computer Music Journal*, **18**, 2, 30–37.

Slawson, W. (1985). *Sound Color*, Berkeley, CA: University of California Press.

Serra, M.-H. (1993). *Stochastic composition and stochastic timbre: GENDY 3 by Iannis Xenakis*, Paris: Cemamu.

Serra, X. and Smith, J. (1990). Spectral modelling synthesis: a sound analysis/synthesis system based on a deterministic plus stochastic decomposition, *Computer Music Journal*, **14**, 4, 12–24.

Smith, J. (1992). Physical modeling using digital waveguides, *Computer Music Journal*, **16**, 4, 74–91.

Stockhausen, K. (1959). ... how time passes ..., *Die Reihe*, 5, English edition, pp. 59–82.

Terhardt, B. (1974). On the perception of periodic sound fluctuation, *Acústica*, **30**, 201–213.

Truax, B. (1988). Real-time granular synthesis with a digital signal processor, *Computer Music Journal*, **12**, 2, 14–26.

Vaggione, H. (1993). Timbre as syntax: a spectral modeling approach, *Contemporary Music Review: Timbre Composition in Electroacoustic Music*, **10**, 2, 73–83.

Wessel, D. (1979). *Timbre space as a musical control structure*, Rapports Ircam, Paris: Centre Georges Pompidou.

Wiener, N. (1964). Spatial–temporal continuity, quantum theory, and music. In Capeck, M. (ed.), *The Concepts of Space and Time*, London: Reidel.

Wishart, T. (1985). *On Sonic Art*, York: Imagineering Press.

Wishart, T. (1994). *Audible Design*, York: Orpheus the Pantomime Ltd.

Wolfram, S. (1986). *Theory and Applications of Cellular Automata*, Singapore: World Scientific.

Xenakis, I. (1971). *Formalized Music*, Bloomington, IN: Indiana University Press.

CD-ROM instructions

ATTENTION:

ALL MATERIALS ON THIS CD-ROM ARE USED AT YOUR OWN RISK.

FOCAL PRESS AND THE AUTHOR OF THIS BOOK CANNOT ACCEPT ANY LIABILITY FOR ANY SOFTWARE MALFUNC-TION OR FOR ANY PROBLEM WHATSOEVER CAUSED BY ANY OF THE SYSTEMS ON THIS CD-ROM.

ANY COMPLAINTS OR ENQUIRIES RELATING TO ANY SYSTEM OR DOCUMENTATION ON THIS CD-ROM MUST BE ADDRESSED DIRECTLY TO THE SOFTWARE AUTHORS AND NOT TO FOCAL PRESS.

In order to run or install a program on your computer, you should normally copy the respective folder onto your hard disk and follow the usual procedures for running or installing software. Most packages provide a Readme file with installation instructions; use Notepad (on PC) or SimpleText (on Mac) to read a Readme file. A few general tips for each program are given below. It is strongly advised that you read them before you go on to use the programs.

As a guidance, the programs are rated as follows:

* = EASY TO GET BY (ideal for the absolute beginner)

** = MEDIUM DIFFICULTY

*** = REQUIRES PRACTICE TO MASTER

PC Software

Tested on a 200 MHz machine running Windows 95 with 1152 × 864 colour display and SoundBlaster AWE64 Gold card.

Also tested on a 166 MHz machine running Windows 95 800 × 600 colour (256) and SoundBlaster AWE64 Gold card.

It is recommended that you use Microsoft Internet Explorer to read the HTML documentation.

1 Audio Architect (in folder Audiarch)*

In order to install the program, double-click on Setup.exe and follow the procedures. You do not need to copy the folder onto your hard disk; Setup runs fine from the CD-ROM. Note, however, that the folder Aanets containing the examples mentioned in the book should be copied manually as this folder is not part of the installable package.

The Quick Start section in the on-line Help is sufficient to allow you to listen to all of the examples. The first network which it suggests you to try is, however, a very long drum sequence which takes several minutes to compute. I suggest that you try shorter examples first, such as fm-fun.mae or groove.mae.

The tutorial in folder Tutorial is a very good way to get started with the program; it is excellent in taking over from where the Beginner's Guide section in the on-line Help finishes.

This is an unregistered demo version of the program but you should be able to make (and save) your own instruments with very few restrictions.

2 CDP (in folder Cdp)**

This package does not need installation as such; simply copy the whole folder onto your hard disk. Note that the CDP programs are designed to run under DOS (command-line interface) rather than Windows. You will need to understand DOS in order to set the paths in the autoexec.bat file as outlined in the documentation.

Bear in mind that output files must **not** already exist and must have the correct extension. Failure to observe these two points will produce an error message: 'failed to open xxx for output'.

Note that sound files and spectral analysis files both use the .wav extension. It might be a good idea to prefix all sound files with 's' and all analysis files with 'a'.

Also, note that the on-line Help for the CDPSPEC program suggests typing 'spec blur' for help with the BLUR process. You should instead type 'cdpspec blur'; this applies for all CDPSPEC processes.

3 Chaosynth (in folder Chsynth)**

This is the author's own program for granular synthesis.

In order to install Chaosynth, double-click on Setup.exe and follow the procedures. You do not need to copy the Chsynth folder onto your hard disk; Setup should run fine from the CD-ROM. Note, however, that the folder Wavfile containing the sound examples mentioned in the tutorial should be copied manually, as this folder is not part of the installable package. Refer to the Readme file for more instructions.

The HTML documentation is far more complete than the on-line Help; use Microsoft Internet Explorer to read the documentation.

4 Generator (in folder Generator)*

In order to install the program, double-click on Gensetup.exe and follow the procedures. You do not need to copy the whole folder onto your hard disk; Setup runs fine from the CD-ROM. Note, however, that the folder Ensembs containing the examples mentioned in the book should be copied manually, as this folder is not part of the installable package.

More information can be found in the Readme.txt file; use Notepad to read this file.

The on-line Help is very useful for running the examples and for designing new instruments. As it is an unregistered demo version, no reference manual is available to describe the functioning of the built-in modules; if you register you will receive one. However, as it is a fairly intuitive piece of software you should be able to figure out how to build complex instruments on a trial and error basis.

5 Morph (in folder Morph)***

This package does not need installation as such; simply copy the whole folder onto your hard disk.

This is a fully working beta version of a program that is still under development. If you contact the author he might send you an upgraded version free of charge.

Morph is a quite complicated package but the manual and the HTML Article give enough information to get the keen reader going.

6 pcmusic (in folder Pcmusic)***

This is a fully working sound synthesis programming language package. It does not need installation as such; simply copy the whole folder onto your hard disk.

The Readme.txt (inside the Pcm folder) gives sufficient information to allow the sample scores to be compiled. However, be aware that the compressed file pcm09.exe is not on the CD-

ROM, as it has already been expanded and all of the files which it would have generated are on the CD-ROM instead.

In folder Text there is a comprehensive tutorial furnished with a complete reference manual specifically compiled for this book by Dr Robert Thompson.

7 Reality (in folder Reality)

No program is provided here. There are, instead, three ScreenCam files that demonstrate the functioning of the program.

8 SMS (in folder Sms)***

This is a fully working program and it does not need installation as such; simply copy the whole folder onto your hard disk.

Article.doc (also saved as postscript file) contains an in-depth explanation of the theory behind the system. Wordpad, MSWord and Word Perfect fail to recognise the format of this file; however, Microsoft Works can open it.

9 Som-A (in folder Soma)**

This is a fully working program and it does not need installation as such; simply copy the whole folder onto your hard disk.

If you repeatedly receive error messages when trying to open/save files, the best procedure is to delete the Soma.cfg file and configure the paths and directories from scratch (refer to the Readme file).

10 Virtual Waves (in folder Virwaves)*

In order to install the program, double-click on Setup.exe and follow the procedures. You do not need to copy the whole folder onto your hard disk; Setup runs fine from the CD-ROM. Note, however, that the folder Synthvir containing the examples mentioned in the book should be copied manually, as this folder is not part of the installable package.

This is an unregistered demo version of the program; some restrictions apply, but all the examples run fine.

11 Wigout and TrikTraks (in folder Wigtrik)***

This is a fully working package. It does not need installation as such; simply copy the whole folder onto your hard disk.

Start by reading the Readme file and then use your Web browser to read the Index.htm file, which is the index for the whole documentation in HTML format. Do not change the configuration of the directories, as this will change the links between the HTML documents.

12 Various

This folder contains three packages: Cool Edit 96, PhyMod and GranuLab.

Cool Edit 96 is not a synthesis program. Rather, it is a sound tool which can be very useful to handle your sound files. You will probably need it at some point (e.g. to convert to different file formats and sampling rates).

PhyMod and GranuLab are two interesting synthesis programs that could not be fully covered by the book, as the manuscript was already in production when these two programs were discovered. They certainly deserve to be featured in the next edition of the book.

GranuLab (in folder Various\Granulab) is a granular synthesis program. It does not require installation; simply copy the folder Granulab onto your hard disk. You will notice that the Patches folder is empty; you should store your own patches in this folder. (rate = *)

PhyMod (in folder Various\Phymod) is a program for physical modelling synthesis. It needs to be unzipped; unzip the file phymod20.zip in your hard disk. A folder containing the whole package is created; no further installation is necessary. (rate = ***).

MACINTOSH Software

Tested on a PowerMac 7100/80W running Mac OS 7.5.5 with 640 × 480 colour display.

1 CLM (in folder Clm)***

This is a complete synthesis programming language package but you will need Common Lisp to run it (MCL and CodeWarrior). You could run CLM with just MCL (i.e. with no CodeWarrior) on a PowerPC Mac running MCL 4.0 or later (earlier versions need CodeWarrior), but it's about 10 times slower than the 'normal' version. In either case, you do need to buy MCL – to date, there is no free Common Lisp version for Macintosh. Other versions such as GCL, CLisp, CMU-CL, or ACL are free, but they do not run on Macs.

As Rhapsody is (or will be) a descendant of Unix/Nextstep, it is hoped that CLisp will run on it (as it runs on the NeXT, for example), providing a free path to Mac platforms.

You should use MSWord to read the Readme.clm file; SimpleText does not work very well here. However, this file may not be entirely relevant at this point because it is assumed that you are downloading the software via the Internet. In fact, you do not need to do this because you already have the system on the CD-ROM – just copy the whole folder onto your hard disk.

There is a user manual in HTML format which can be read by a standard Web browser: clm.html. Also, there is a very good tutorial in folder tutorial: toc.html.

2 Diphone (in folder Diphone)***

Diphone must be installed on your computer. You can either install it from the CD-ROM or copy the folder onto your hard drive prior to installation. You will need to understand basic French in order to extract the files and put them in the right places.

As the 'Put in Extension Folder' directory is empty, you should bear in mind that the actual folder that you must put in the Extension Folder of your Mac is 'Ircam Kernels 1.7.1'.

You will need a PDF reader to read the Reference Manual and an AIFF player to play the sounds.

Contact Ircam in Paris for more information on their range of Mac software for computer music applications.

3 GrainWave (in folder GrainWave)*

This is a fully working package; but you should register as an act of goodwill, as the programmer worked very hard to make this a unique Mac program of its kind.

It does not need installation as such; simply copy the whole folder onto your hard disk. The tutorial is easy to follow and provides a good introduction to the package.

4 LASy***

This is a fully working package. It does not need installation as such; simply copy the whole folder onto your hard disk.

You need MSWord to read the documentation. Beginners may find it slightly difficult to follow, largely due to the complexity of the software and theoretical nature of the text.

Enjoy!

Eduardo Reck Miranda

Index

Low-pass filters (LPF), 97–8

McAlpine, Kenny, 30, 75, 126
Machine code, 8
Machine-learning techniques, 170
Macro-modules, *see* Subroutines
Maintenance process, sound
 development, 150
Manning, Peter, 182
Mathews, Max, 8
Max Virtual Controller, 36–7
Memory types, 177
MIDI-based control, 15–16
Miller, Joan, 8
MIMD (Multiple Instructions
 Multiple Data) machines,
 177–9
Modal synthesis technique, 95–6
Modelling techniques (synthesis):
 loose, *see* Loose modelling
  techniques
 physical, *see* Physical modelling
  techniques
 spectral, *see* Spectral modelling
  techniques
 time, *see* Time modelling
  techniques
Modulation index, 60, 64, 68, 69
Modulator (audio signal), 58, 61
Moore, F. Richard, 9, 18
MORPH (software synthesis),
 54–6, 138, 139
Mosaic synthesis system, 96
Music N (programming
 languages), 8–9
 classic type, 18
 instrument files, 15–16

Native Instruments (company), 30
Negative amplitudes, 67
Negative frequencies, 68
NeXT/Sun (sound file format), 6
Noise source, 155
Non-linear distortion, *see*
 Waveshaping synthesis
Non-linear processing, *see*
 Waveshaping synthesis
Non-standard synthesis, *see* Binary
 instruction synthesis;
 Sequential waveform
 composition
Normal voicing, 154
Note lists specification, pcmusic,
 20–1
Nyquist distortion, 68

Nyquist frequency/limit, 5

Olson, H. F., 91
OMIT program, 118
Open MIDI System (OMS), 36
Operators (higher level modules),
 74
Orchestra file, Music N software, 15
Oscil/oscillator, 8, 14, 15
Overlap-add resynthesis, 135

Parallel computing, 176
 architectures, 176–9
 strategies, 179–81
Partials:
 amplitude calculation, 64–6
 frequency calculation, 63–4
Patch (algorithm), 75
Path selection, programming,
 10–12
Pcmusic (programming language),
 9, 18–19
 additive synthesis examples,
  126
 amplitude modulation
  examples, 61
 codification example, 14–15
 complete score example, 21–2
 instrument code, 157–60
 instrument specifications, 19–20
 jitter generator code, 152
 Karplus–Strong example, 103
 lookup table specifications, 20
 noise source code, 155
 note list specifications, 20–1
 voicing source code, 154
Peak-tracking algorithm, 136
Perceptual attributes,
 generalisation, 169–70
Phase Vocoder system, 136–8
PhyMod program, 93
Physical modelling (PM)
 techniques, 90–2
 classic approach, 92–3
 commercial synthesisers, 92
 functional approach, 93–4
 implementation, 95
 Karplus–Strong, 102–3
 modal synthesis, 95–6
 recirculating wavetable
  approach, 95
 subtractive synthesis, 96–100
 waveguide filtering, 100–2
PILE (binary instruction
 synthesis), 83

Polynomials, for spectra
 prediction, 79–80
Predictive analysis, 132–3
Procedures, *see* Subroutines
Processing elements, 176–7
Program translator, 8
Programming languages, 8–9, *see
 also under specific names*
Programming systems:
 Audio Architect, 28–30
 GENERATOR, 30–2
 GrainWave, 35–8
 Virtual Waves, 32–5

Quantisation noise, 4
Quasi-sinusoidal source signal,
 154

Rayleigh, J. W. S., 91
Reality (software synthesiser),
 38–40
RESHAPE program, 118
Resynthesis:
 by Fourier analysis reversion,
  135–6
 by fragmentation and growth,
  114–15
 by reintegration of discarded
  components, 140, 142
 in spectral modelling, 133–43
Ring Modulation (RM), 60, 131
Risset, Jean-Claude, 76
Roads, Curtis, 107
Rodet, Xavier, 42, 144

S/PDIF (sound transfer format), 7
Sampling process, 3–4
Sampling theorem, 4
Scaling factors, 65–6
 table, 186
Schaeffer, Pierre, 124, 150
Schottstaedt, Bill, 9, 22
Score file, Music N software, 15
Seer Systems (company), 38
Sequential waveform composition,
 120–3
Serra, Xavier, 50, 142
Short-time Fourier transform
 (STFT), 127–9
 wavelets analysis, 132
 windowing, 129–32
SHUFFLE program, 138
Signal modifiers, 14
SIMD (Single Instruction Multiple
 Data) machines, 176–7